FUNDAMENTAL AND APPLIED PROBLEMS OF TERAHERTZ DEVICES AND TECHNOLOGIES

Selected papers from the Russia-Japan-USA-Europe Symposium (RJUSE-TeraTech 2016)

SELECTED TOPICS IN ELECTRONICS AND SYSTEMS

Editor-in-Chief: **M. S. Shur**

---

*The complete list of the published volumes in the series can be found at
http://www.worldscientific.com/series/stes

Selected Topics in Electronics and Systems – Vol. 58

FUNDAMENTAL AND APPLIED PROBLEMS OF TERAHERTZ DEVICES AND TECHNOLOGIES

Selected papers from the Russia-Japan-USA-Europe Symposium (RJUSE-TeraTech 2016)

Katahira Campus of Tohoku University, Sendai, Japan
31 October – 4 November 2016

Editors

M. Ryzhii
A. Satou
T. Otsuji
University of Aizu, Japan

W⊖ World Scientific

NEW JERSEY · LONDON · SINGAPORE · BEIJING · SHANGHAI · HONG KONG · TAIPEI · CHENNAI · TOKYO

Published by

World Scientific Publishing Co. Pte. Ltd.
5 Toh Tuck Link, Singapore 596224
USA office: 27 Warren Street, Suite 401-402, Hackensack, NJ 07601
UK office: 57 Shelton Street, Covent Garden, London WC2H 9HE

British Library Cataloguing-in-Publication Data
A catalogue record for this book is available from the British Library.

Selected Topics in Electronics and Systems — Vol. 58
**FUNDAMENTAL AND APPLIED PROBLEMS OF TERAHERTZ DEVICES
AND TECHNOLOGIES**
Selected Papers from the Russia-Japan-USA-Europe Symposium (RJUSE-TeraTech 2016)

Copyright © 2017 by World Scientific Publishing Co. Pte. Ltd.

ISBN 978-981-3223-27-1

Preface

Terahertz (THz) electromagnetic waves, phenomena in the THz range and related technological issues have been explosively investigated during the recent two decades, its potential as a disruptive technology to commercial applications has yet to make any impression.

The Russia-Japan-USA-Europe Symposium on Fundamental and Applied Problems of Terahertz Devices and Technologies (RJUSE-TeraTech 2016) took place at Katahira Campus of Tohoku University, Sendai, Japan on October 31 – November 4, 2016 and was the fifth in the series of preceding successful symposiums held in Japan, Russia, and USA. The symposium aims to bring together researchers who tackle the broad range of related problems in the terahertz devices, technologies and applications, so as to stimulate discussions on their state-of-the-art results and promote their collaborations.

This issue contains selected extended papers presented as invited and contributed talks at the RJUSE-TeraTech 2016 symposium. It addresses the variety of topics, in particular, THz detectors based on double heterojunction bipolar transistors (DHBT) and field effect transistors (FET) utilizing resonant plasma effects, quantum cascade (QCL) and HgCdTe quantum-well heterostructures, and graphene-based THz devices.

We thank the following sponsors for their financial contributions to the success of this symposium — the Research Institute of Electrical Communication, Inoue Foundation for Science, Intelligent Cosmos Scientific Foundation, the Murata Science Foundation, and the Foundation for the Promotion of Electrical Communication Engineering.

M. Ryzhii, A. Satou, and T. Otsuji

The Editors
RJUSE-TeraTech 2016
The Russia-Japan-USA-Europe Symposium

Contents

High-Speed Room Temperature Terahertz Detectors Based on InP Double Heterojunction Bipolar Transistors

D. Coquillat[1,*], V. Nodjiadjim[2], S. Blin[3], A. Konczykowska[2], N. Dyakonova[1], C. Consejo[1], P. Nouvel[3], A. Pénarier[3], J. Torres[3], D. But[1], S. Ruffenach[2], F. Teppe[1], M. Riet[2], A. Muraviev[4], A. Gutin[4], M. Shur[4], and W. Knap[1]

[1]*Laboratoire Charles Coulomb (L2C), UMR 5221 CNRS-Université de Montpellier, Montpellier, FR-34095, France*

[2]*III-V Lab, Campus de Polytechnique, 1 avenue Augustin Fresnel, Palaiseau, FR-91767, France*

[3]*Institut d'Electronique et des Systèmes, UMR 5214 CNRS-Université de Montpellier, Montpellier, FR-34095, France*

[4]*Rensselaer Polytechnic Institute, Troy, New York 12180, United States of America*
dominique.coquillat@umontpellier.fr

Compact and fast detectors, for imaging and wireless communication applications, require efficient rectification of electromagnetic radiation with frequencies approaching 1 THz and modulation bandwidth up to a few tens of GHz. This can be obtained only by using a mature technology allowing monolithic integration of detectors with low-noise amplifiers. One of the best candidates is indium phosphide bipolar transistor (InP HBT) technology. In this work, we report on room temperature high sensitivity terahertz detection by InP double-heterojunction bipolar transistors (DHBTs) operating in a large frequency range (0.25–3.1 THz). The performances of the DHBTs as terahertz sensors for communications were evaluated showing the modulation bandwidth of investigated DHBTs close to 10 GHz.

Keywords: Double heterojunction bipolar transistor (DHBT); THz power detectors; THz imaging.

1. Introduction

Heterojunction Bipolar transistors (HBTs) have emerged as promising devices for sensitive and room temperature detection of terahertz (THz) radiation. They are considered as a good candidate for imaging, tomography, spectroscopy, as well as for future high data-rate THz wireless communications.[1-6] These applications require high speed and high responsivity detectors, integrated monolithically with antenna and read-out electronics.

The advanced technologies suitable for these high speed applications are only available in: *i)* high-electron mobility transistors (HEMTs) and *ii)* InP heterojunction bipolar transistors (HBTs). InP HEMTs were actively investigated in the past and they were demonstrated as THz detectors with extremely good sensitivity.[7-9] There exist only a few reports on InP HBT based THz detectors.[4,10-12]

In this work, we report on a THz responsivity study of InP double-heterojunction bipolar transistors (DHBTs) and demonstrate that these transistors can be competitive sensitive THz sensors operating in a large radiation frequency range (0.25–3.1 THz) with the modulation bandwidth up to 10 GHz. We present preliminary characterization of these detectors for high-data rate wireless communications at carrier frequency close to 300 GHz atmospheric window.

2. Device Structure and Fabrication

The 0.7x5 μm^2 emitter size InP/InGaAs DHBTs used in our experiments were designed for high bit-rate optical communications applications.[13] They were fabricated using self-aligned triple-mesa process for patterning the emitter, base and collector [Fig. 1(a)], and demonstrated current gain cutoff frequency (f_T) and maximum oscillation frequency (f_{max}) above 320 and 280 GHz, respectively. The static current gain was around 40 and the common-emitter breakdown voltage was above 5V. As shown by a micrograph of the device [Fig. 1(b)], they operated without any specific spatial coupling antennas. The radiation was coupled to the device only via the contact pads and/or the coupling wires to the device [Fig. 1(c)].

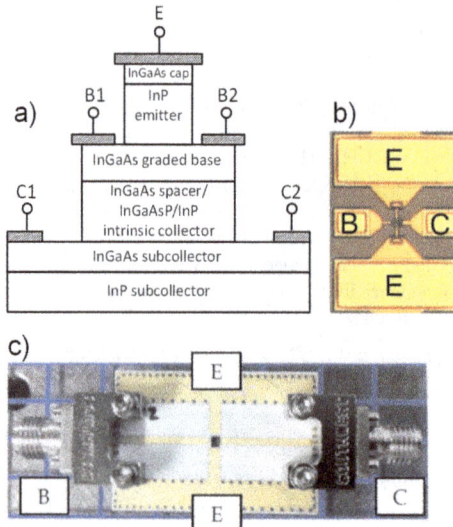

Fig. 1. (a) Device structure, (b) micrograph of the DHBT and configuration of the contact pads. (c) Package of the test chip with DHBT mounted in the middle.

3. Photoresponse at Low Frequency Modulated THz Signal

For the low frequency modulated photoresponse measurements we used electronic tunable sources based on frequency multipliers covering the ranges from 270 to 690 GHz and a THz continuous-wave CO_2 - pumped molecular gas laser with frequencies in the range 0.91–3.11 THz. The incoming THz radiation was chopped at low frequency in the range 100–650 Hz. The detectors can be used in two basic configurations: *i)* a passive configuration for which the base-emitter junction nonlinearity is used as THz rectifier, while the collector terminal is left open, and *ii)* an active configuration for which the DHBT is also biased at the collector port. In this work we study only the passive one.

To characterize the detectors behavior in the passive configuration, the V_{BE} bias dependence of the photovoltage was read at the detector collector terminal using a synchronized lock-in amplifier. The emitter terminal of the device was grounded. Prior to photoresponse investigation, I_{BE}-V_{BE} characteristics were performed when the collector terminal was left open. In this case, V_{BE} was controlled and I_{BE} measured using a Keithley Source Meter. The induced photocurrent ΔI_{BE}, superimposed on I_{BE}, was also measured as the difference of the characteristics with and without applied THz radiation.

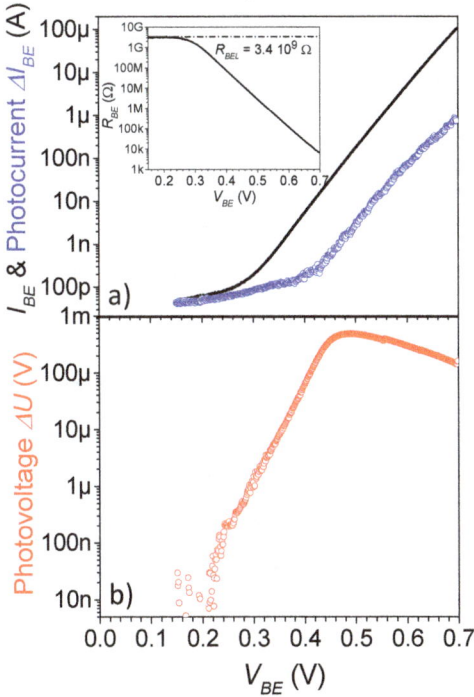

Fig. 2. (a) I_{BE}-V_{BE} characteristics (solid black line) and induced photocurrent ΔI_{BE} (open blue dots) superimposed on I_{BE} and measured as the difference of the characteristics with and without applied 286-GHz radiation. Inset: measured resistance between base and emitter terminals. The intrinsic DHBT static resistance is in parallel with the pad resistance R_{BEL}. (b) Photovoltage in passive configuration (the base-emitter junction nonlinearity is used for detection, while the collector is used to collect the *dc* down-converted signal) for 286 GHz radiation (open red dots).

Figure 2(a) shows both the I_{BE}-V_{BE} characteristics for one of the test DHBTs, and the photocurrent ΔI_{BE}. The leakage currents observed in the low V_{BE} range are associated with parasitic current paths between measurement pads and do not concern the intrinsic DHBT.[14] The variation of the measured resistance between base and emitter terminals is shown in the inset of Fig. 2(a). The value of the pad resistance R_{BEL}, in parallel with the intrinsic HBT resistance and responsible for the leakage current evolution was extracted.[14] Figure 2(b) shows an example of typical experimental photovoltage ΔU measured at the collector terminal at 286 GHz, as a function of V_{BE}. One can see that with the increase of V_{BE}, the amplitude of the photovoltage achieves its maximum around 0.49 V and drops down. In contrast, the induced photocurrent ΔI_{BE} increases monotonically up to 0.70 V.

The detector circuit makes use of the base-emitter junction as a nonlinear passive element to rectify the *ac* THz signal. The collector is used to collect the *dc* signal from the base region. In the passive configuration, if the load impedance of the read-out circuit is much greater than the static resistance the small-signal photovoltage, ΔU can be described from the current-voltage characteristics as[16]

$$\Delta U = A \frac{d}{dV_{BE}} ln I_{BE}(V_{BE}) \tag{1}$$

where A is the parameter reflecting the coupling efficiency and the power of the THz source.

Figure 3 shows the comparison of the measured photovoltage (red open dots) and the calculated photovoltage using Eq. (1) (open gray triangles). The experimental photovoltage in Fig. 2 can be explained by this model only for relatively large V_{BE} bias. At low V_{BE}, the experimentally observed photovoltage is suppressed and this feature cannot be explained by Eq. (1).

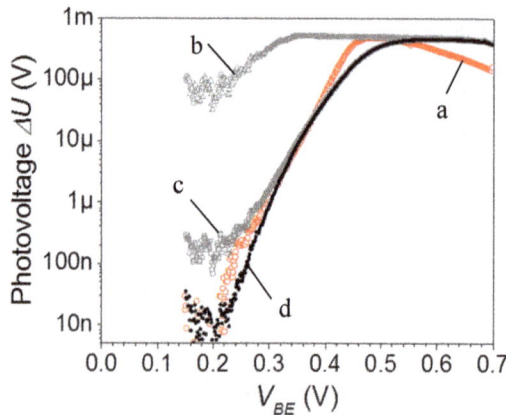

Fig. 3. Comparison of directly measured photovoltage (open red dots, a) and calculated photovoltage as a function of V_{BE} from the I_{BE}-V_{BE} characteristic by Eq. (1) (open gray triangles, b), by Eq. (2) when taking into account the loading effect (open gray squares, c), and for the combined effect of the loading effect and the leakage current (full black dots, d). The fitting parameter used for the three calculated curves was $A = 1.42 \times 10^{-5}$.

Two separate contributions can be responsible of the attenuation of the responsivity for the low V_{BE} range: the loading effect related to capacitive and resistive coupling of the

DHBT detector to the read-out circuit, and the leakage current between the pads surrounding the intrinsic DHBT.[14] The analyze of the loading effect in the case of field effect transistors (with resistance R_{det}) using the equivalent circuit of the detector and loaded with the measurement setup (cables, connectors, preamplifier) has allowed for an accurate description of the experimental photovoltage.[15,16] Using a simple voltage divider model, the photovoltage can be simulated with the following expression:

$$\Delta U_{fit} = A \frac{d}{dV_{BE}} ln I_{BE}(V_{BE}) \frac{1}{1+R_{det}/Z_L} \tag{2}$$

where Z_L is the load impedance of the set up (with input resistance of the preamplifier 10 MΩ, capacitance of bounding wires, cables and connectors and bounding wires 20 pF, chopper frequency 619 Hz). Figure 3 shows the calculated photovoltage taking into account the loading effect (pen gray squares). As expected, the loading effect results in dramatically reduced photovoltage when the resistance R_{BE} of the emitter-base junction becomes larger (at $V_{BE} < 0.46$ V) than the load impedance Z_L of the output circuitry.

When V_{BE} becomes smaller than 0.25 V, the leakage current from the parasitic paths between the measurement pads is predominant [see Fig. 2(a)]. To take into account the leakage current, the intrinsic emitter-base resistance for the low V_{BE} range can be computed using the value of the pad resistance R_{BEL} and used to recalculate the shape of the photovoltage defined by Eq. (2). The recalculated photovoltage (full black dots) shows a better description of the measured one. These preliminary results allow us to understand the behavior of the photovoltage and photocurrent measured in the passive configuration.

In Fig. 4 we have plotted the device area-normalized voltage responsivity R_V obtained between the collector-emitter electrodes of the 0.7-μm DHBT as a function of base-emitter bias V_{BE} for 335 GHz and for four higher frequencies. The responsivity R_V is extracted from the measured ΔU by using the relation $R_V = (\Delta U S_t)/(P_t S_a)$, where P_t is the power of the incoming radiation in the focal plane of the device, $S_t = \pi d^2/4$ is the radiation beam spot area, and S_a is the total area of the device which includes contact pads (280 μm x 385 μm). As observed in Fig. 4, the device area normalized responsivity degrades with increasing frequency.

The log-log plot of the responsivity maxima is shown in the inset of Fig. 4. It decreases approximately as ω^{-6}, where ω is the angular frequency of the incident wave. There is no strict quantitative theory explaining this roll off. We believe that it results from two effects: *i)* the parasitic capacitances and inductances[17] and *ii)* the fact that the metallized pads (playing the role of antennas) are placed on thick dielectric substrate introducing radiation losses related to the substrate modes.[18,19] The effects of the parasitic elements were studied in Ref. 17, taking into account the combined effects of the capacitance, C, shunting the active nonlinear device resistance, R, and parasitic series inductance, L. The voltage drop V_R across R becomes independent on R and proportional to $1/(\omega^2 LC)$. The detector response considering the parasitic elements is proportional to V_R^2, and, therefore inversely proportional to $1/(\omega^4 L^2 C^2)$. In addition, due to the influence of the substrate modes, a significant portion of incoming radiation is coupled to the substrate instead of being coupled to the transistor, leading to responsivity losses. Theoretical investigations of the responsivity versus substrate thickness were performed, notably for common materials used in semiconductor technologies, including InP substrates.[19] When the radiation frequency increases, the amount of power trapped in the dielectric substrate rapidly grows. This gives an additional frequency dependant factor that contributes to the degradation of the responsivity at higher frequencies.

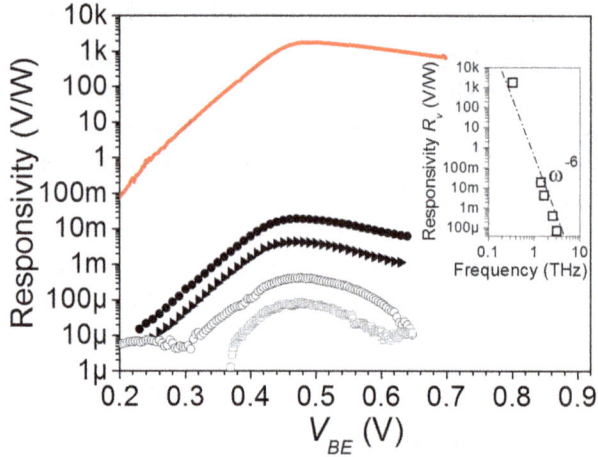

Fig. 4. Area-normalized responsivity R_V as a function of the base-emitter bias V_{BE} for frequency of the incident THz radiation of 335 GHz (solid red curve), 1.40 THz (●●●), 1.63 THz (▲▲▲), 2.52 THz (OOO), and 3.11 THz (□□□). Inset: log-log plot of the responsivity maxima around $V_{BE} = 0.49$ V as a function of the radiation frequency.

4. Modulation Bandwidth

Wireless high-data-rate error-free communications have been already demonstrated up to 8.3 Gbps using transistors as THz detectors,[5,6] but for high-electron mobility transistors only. As already mentioned, transistors are very attractive for their integration with monolithic high-speed integrated circuits, mainly due to their tunable and low output impedance. Here, we investigate the performances of InP DHBT as THz detectors for communications, by studying their output modulation bandwidth. We used a commercial frequency-multiplied source whose last stage is a mixer, in order to offer a carrier-suppressed amplitude modulation up to 30 GHz. We applied a sine-wave modulation signal at frequency f_m. The THz beam was collimated using a Teflon lens whose focal length was 10 cm, then focused using the same lens on the transistor. We study output modulation bandwidth of DHBT biased at $V_{BE} = 0.7$ V (without collector–emitter bias). Since the DHBT is a non-linear detector (as it is sensitive to the incoming radiation power) excited by a carrier-suppressed modulated signal, we measure a collector-emitter signal at two-times f_m.

As shown in Fig. 5, we observe a wide modulation bandwidth with cut-off frequency close to 10 GHz mainly limited by the bonding wires. We interpret the strong antiresonances as due to the impedance mismatch of the DHBT and the transmission line. It can be improved in the future by using active detector configuration with an adequate biasing between collector and emitter allows to control/reduce DHBT output impedance.

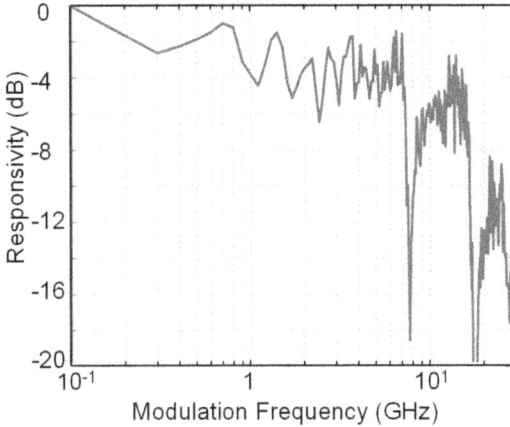

Fig. 5. Dependence of sensitivity on modulation frequency for a 295-GHz carrier frequency and for $V_{BE} = 0.70$ V. The reported modulation frequency corresponds to the frequency of the detected signal.

5. Conclusions

We evaluate the optical performance of the InP DHBTs as room temperature detectors in the passive configuration for a large frequency range – up to 3 THz. We show that they can operate far above the frequencies at which they have gain and can still rectify THz voltage and current. The maximum of photovoltage as a function of the emitter-base bias V_{BE} can be explained by the combined effect of the loading effect related to capacitive and resistive coupling of the DHBT detector to the read-out circuit and the leakage current between the measurements pads surrounding the intrinsic transistor. The performances of the DHBTs as THz detectors for communications, in the passive configuration, were evaluated at 300 GHz carrier frequency, by studying their output modulation bandwidth. We demonstrated that already in this preliminary study the modulation bandwidth of DHBT can go up to 10 GHz. We predict that the modulation bandwidth can be still improved by use of high frequency support/connectors and by use of bias between collector and emitter (active configuration) in order to reduce DHBT output impedance. Therefore DHBT can be very interesting not only for fast THz imaging but also for future high-data rate wireless communication applications.

Acknowledgments

This work was financially supported by the ANR project NADIA "Integrated NAno-Detectors for terahertz Applications" (ANR-13-NANO-0008), by the financial support from ERA.NET RUS Plus through the project Terasens n°149, by the Region of

Occitanie/Pyrénées-Méditerranée through the GEPETO "Terapole Platform" and HERMES-Obi-@ Platform.

References

1. R. A. Hadi, J. Grzyb, B. Heinemann, and U. Pfeiffer, "Terahertz detector arrays in a high-performance SiGe HBT technology", *2012 IEEE Bipolar/BiCMOS Circuits and Technology Meeting (BCTM)*, Portland, OR, USA, Oct. 2012.
2. D. Yoon, J. Yun, J. Kim, K. Song, M. Kaynak, B. Tillack, and J.-S. Rieh, "3-D THz tomography with an InP HBT signal source and a SiGe HBT imaging receiver operating near 300 GHz", *Proc. 40th IRMMW-THz: Int. Conf. on Infrared, Millimeter, and Terahertz Waves (IRMMW-THz)*, Hong-Kong, China, Aug. 2015.
3. S. Carpenter, D. Nopchinda, M. Abbasi, Z. He, M. Bao, T. Eriksson, and H. Zirath, "A *D*-band 48-Gbit/s 64-QAM/QPSK direct-conversion I/Q transceiver chipset", *IEEE Trans. Microw. Theory Tech.* **64** (2016) 1285-1296.
4. D. Coquillat, V. Nodjiadjim, A. Konczykowska, N. Dyakonova, C. Consejo, S. Ruffenach, F. Teppe, M. Riet, A. Muraviev, A. Gutin, M. Shur, J. Godin, and W. Knap, "InP double heterojunction bipolar transistor for broadband terahertz detection and imaging systems", *J. Phys.: Conf. Ser.* **647(1)** (2015) 012036.
5. S. Blin, L. Tohme, D. Coquillat, S. Horiguchi, Y. Minamikata, S. Hisatake, P. Nouvel, T. Cohen, A. Penarier, F. Cano, L. Varani, W. Knap, T. and Nagatsuma, "Wireless communication at 310 GHz using GaAs high-electron-mobility transistors for detection", *J. Commun. Networks* **15(6)** (2013) 559-568.
6. L. Tohmé, S. Blin, G. Ducournau, P. Nouvel, D. Coquillat, S. Hisatake, T. Nagatsuma, A. Pénarier, L. Varani, W. Knap, and J. F. Lampin, "Terahertz wireless communication using GaAs transistors as detectors", *Electron. Lett.* **50** (2014) 323-325.
7. V. V. Popov, D. V. Fateev, T. Otsuji, Y. M. Meziani, D. Coquillat, and W. Knap, "Plasmonic terahertz detection by a double-grating-gate field-effect transistor structure with an asymmetric unit cell", *Appl. Phys. Lett.* **99** (2011) 243504.
8. Y. Kurita, G. Ducournau, D. Coquillat, A. Satou, K. Kobayashi, S. Boubanga Tombet, Y. Meziani, V. V. Popov, W. Knap, T. Suemitsu, and T. Otsuji, "Ultrahigh sensitive sub-terahertz detection by InP-based asymmetric dual-grating-gate high-electron-mobility transistors and their broadband characteristics", *Appl. Phys. Lett.* **104** (2014) 251114.
9. P. Nouvel, J. Torres, S. Blin, H. Marinchio, T. Laurent, C. Palermo, L. Varani, P. Shiktorov, E. Starikov, V. Gruzinskis, F. Teppe, Y. Roelens, A. Shchepetov, and S. Bollaert, "Terahertz emission induced by optical beating in nanometer-length field-effect transistors", *J. Appl. Phys.* **111** (2012) 1037072.
10. D. Coquillat, V. Nodjiadjim, A. Konczykowska, M. Riet, N. Dyakonova, C. Consejo, F. Teppe, J. Godin, and W. Knap, "InP double heterojunction bipolar transistor as sub-terahertz detector", *Proc. 2014 39th Int. Conf. on Infrared, Millimeter, and Terahertz Waves (IRMMW-THz)*, Tucson, AZ, USA, Sep. 2014.
11. V. Vassilev, H. Zirath, R. Kozhuharov, and S. Lai, "140–220-GHz DHBT detectors", *IEEE Trans. Microw. Theory Tech.* **61(6)** (2013) 2353-2360.
12. V. Vassilev, N. Wadefalk, R. Kozhuharov, M. Abbasi, S. E. Gunnarsson, H. Zirath, T. Pellikka, A. Emrich, M. Pantaleev, I. Kallfass, and A. Leuther, "MMIC based components for MM-Wave instrumentation", *IEEE Microw. Wireless Comp. Lett.* **20(10)** (2010) 578-580.
13. J.-Y. Dupuy, A. Konczykowska, F. Jorge, M. Riet, P. Berdaguer, V. Nodjiadjim, J. Godin, and A. Ouslimani, "A large-swing 112-Gb/s selector-driver based on a differential distributed amplifier in InP DHBT technology", *IEEE Trans. Microw. Theory Tech.* **61** (2013) 517-524.

14. J. C. Martin, C. Maneux, N. Labat, A. Touboul, M. Riet, S. Blayac, M. Kahn, and J. Godin, "Extrinsic leakage current on InP/InGaAs DHBTs", *Int. Conf. on Indium Phosphide and Related Materials*, 2003, Santa Barbara, CA, USA, May 2003, pp. 12-15.

15. W. Stillman, M. S. Shur, D. Veksler, S. Rumyantsev, and F. Guarin, "Device loading effects on nonresonant detection of terahertz radiation by silicon MOSFETs", *Electron. Lett.* **43(7)** (2007) 422-423.

16. M. Sakowicz, M. Lifshits, O. Klimenko, F. Schuster, D. Coquillat, F. Teppe, and W. Knap, "Terahertz responsivity of field effect transistors versus their static channel conductivity and loading effects", *J. Appl. Phys.* **110** (2011) 054512.

17. A. Gutin, T. Ytterdal, A. Muraviev, and M. Shur, "Modelling effect of parasitics in plasmonic FETs", *Solid-State Electron.* **104** (2015) 75-78.

18. P. Kopyt, P. Zagrajek, J. Marczewski, K. Kucharski, B. Salski, J. Lusakowski, W. Knap, and W. K. Gwarek, "Analysis of sub-THz radiation detector built of planar antenna integrated with MOSFET", *Microelectron. J.* **45(9)** (2014) 1168-1176.

19. D. Coquillat, J. Marczewski, P. Kopyt, N. Dyakonova, B. Giffard, and W. Knap, "Improvement of terahertz field effect transistor detectors by substrate thinning and radiation losses reduction", *Opt. Exp.* **24(1)** (2016) 272-281.

Terahertz Response of Tightly Concatenated Two Dimensional InGaAs Field-Effect Transistors Integrated on a Single Chip

D. M. Yermolayev[*1], E. A. Polushkin[2], A. V. Koval'chuk[3], and S. Yu. Shapoval[4]

Laboratory of Epitaxial Micro- and Nanostructures,
Institute of Microelectronic Technology and High-Purity Materials, Chernogolovka, 142432, Russia
[*]JSC "Russian Space System", Moscow, 111250, Russia
[1]*Yermolayev.Denis@gmail.com*
[2]*evgeny@iptm.ru*
[3]*anatoly-fizmat@mail.ru*
[4]*shapoval@iptm.ru*

K. V. Marem'yanin[5] and V. I. Gavrilenko[6]

Department for Physics of Semiconductors, Institute for Physics of Microstructures,
Nizhny Novgorod, 603950, Russia
[5]*kirillm@ipmras.ru*
[6]*gavr@ipmras.ru*

N. A. Maleev[7] and V. M. Ustinov[8]

Physics of Semiconductor Heterostructure, Ioffe Physical Technical Institute,
St. Petersburg, 194021, Russia
[7]*maleev.beam@mail.ioffe.ru*
[8]*vmust@beam.ioffe.ru*

V. E. Zemlyakov[9], V. I. Yegorkin[10], and V. A. Bespalov[11]

National Research University of Electronic Technology, Zelenograd, 124498, Russia
[9]*vzml@rambler.ru*
[10]*egorkin@qdn.miee.ru*
[11]*vrfin@miee.ru*

V. V. Popov

Kotelnikov Institute of Radio Engineering and Electronics (Saratov Branch), Saratov, 410019, Russia
Physics Department, Saratov State University, Saratov, 410012, Russia
Saratov Scientific Center of the Russian Academy of Sciences, Saratov, 410028, Russia
popov_slava@yahoo.co.uk

I. Khmyrova

University of Aizu, Aizu-Wakamatsu, Aizu-Wakamatsu, 965-8580, Japan
khmyrova@u-aizu.ac.jp

THz response of AlGaAs/InGaAs/GaAs HEMT structure has been investigated. The structure consists of the serpentine chain of series connected HEMTs. The source of one is the drain for the subsequent transistor. Experiments have been showed THz response peculiarities of such structures and enhanced noise equivalent power.

Keywords: terahertz radiation; transistor; detector; plasmon; gallium arsenide; grating.

1. Introduction

Recent research in solid-state electronic devices for terahertz radiation (THz) detection has been demonstrated parameters suitable for practical applications, for example, such important parameters as a responsivity and noise equivalent power (NEP).[1–6] There are two main types of designs of detectors: transistor with the antenna[2,3,6] and chip/transistor from the combination of ohmic and potential contacts.[1,4,5] Most of highly sensitive detectors has lateral dimensions several times smaller than a half wavelength of THz[7] radiation and experiments are conducted on a single detector surrounded only by contact wiring. The traditional quasi-optical systems can focus a beam of THz radiation to a spot diameter of a half-wavelength that will overlap several of the detectors in the focal-plane array (FPA). Therefore the radiation power will be distributed between the neighboring detectors but also do not forget about the wiring that is different in the cases of one detector and FPA. Thus a single detector in the FPA will share the power with the other detectors in FPA and the responsivity of the detector would be diminished. Therefore it is necessary to study the behavior of detectors in the FPA or to process a single detector with dimensions of half-wavelengths and then investigate its properties.

In Ref. 6, transistor with the antenna design (patch antenna) and integrated amplifier were used. The size of pixel and half-wavelength of the radiation are 200×150 μm^2 and 230 μm respectively with 3×5 pixels in FPA. The detectors have the following performance data at room temperature: responsivity 70 kV/W with 43 dB voltage amplifier at 0.65 THz, NEP 300 $pW/\sqrt{Hz}$ at 30 kHz. In **Ref. 8**, the authors investigated a single detector with a log-periodic aerial antenna with diameter 1.5 mm at frequency range 200–600 GHz. This detector has performance data at liquid helium temperature: a responsivity of 2.5 kV/W with 240 GHz excitation, calculated NEP ~ 100 $pW/\sqrt{Hz}$ at 750 Hz. At the same time there are many works where it has been studied differently configured transistors on a single chip,[1,4,5] but were not represented FPAs or single detectors with a size corresponding half-wavelength of radiation of such design.

In this paper, the experimental results are presented on the THz response of the detector which is considered as the FPA with serial connected single transistors or as the single detector with dimensions corresponding half-wavelength of radiation.

2. Fabrication and Measurements

2.1. *Fabrication*

A material system for the devices is an AlGaAs/InGaAs/GaAs system grown by molecular beam epitaxy. The two-dimensional (2D) electron channel was formed in 12-nm-thick undoped InGaAs layer with 40-nm-thick δ-doped AlGaAs carrier-supplying layer, and 400-nm-thick undoped GaAs buffer layer formed on the (100) surface of 450-nm-thick semi-insulating (SI) GaAs substrate. 50-nm-thick cap GaAs layer was n-doped with Si up to 6×10^{18} cm^{-3}. The electron density in the channel is 3×10^{12} cm^{-2} with the electron effective mass $m^* = 0.061 m_0$, where m_0 is the free-electron mass, and room temperature mobility is 5 900 cm^2/V·s (which corresponds to the electron scattering time $\tau = 0.2$ ps).

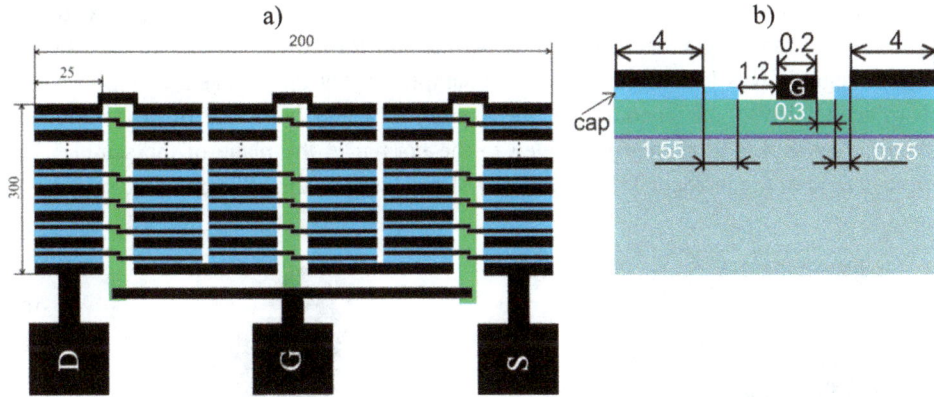

Fig. 1. (a) Top view of the detector (dimension in μm). (b) The vertical structure of single transistor in the array (dimension in μm).

The array of identical FETs, all of which were connected in series, was fabricated on a single chip [Fig. 1(a)]. An individual FET is schematically shown in Fig. 1(b). The source and drain ohmic contacts were made by deposition of 80-nm-thick AuGe-Ni-Au (30/10/40 nm) layer with further annealing at 400°C for 30 s in nitrogen ambient. Gate metallization (metal-semiconductor contact) was made by 150/500 nm Ti-Au layers. The substrate was thinned to 100 μm. All metal layers were deposited by e-beam evaporation. The metal contact patterning was performed by the e-beam lithography and a standard lift-off process. The top surface of the sample was passivated by deposition of thin silicon nitride layer. Gate bus is connected with the metal gate strips via resistors made of a mesa to avoid shunting THz induced current between metal gate strips via metal connections.[9] In addition, when moving to the next line of transistors it is changed the symmetry of the location of the metal strips of the gate which changes the direction of the rectified current. This decision allows flowing the rectified current in every line of transistors in one direction. The detector has 6 lines of transistors with 37 transistors in every one. The dimension of the detection area is 300×200 μm². The channel width of transistors in lines is 25 μm.

2.2. *Measurement setup*

The crystal was mounted in the sample holder allowing for applying dc gate bias to the common gate pad and dc drain bias between the side (source and drain) pads of the transistor. Two monochromatic source of radiation have been used: backward wave oscillator (BWO) and RF generator. BWO was used as a source of sub-THz radiation in frequency range 415–720 GHz with output power equal to 1 mW. The RF generator has been set for one frequency 143 GHz with its maxima power of 50 mW. Power attenuation in the oversized circular copper waveguide was about 10 dB. Terahertz radiation from the BWO was mechanically chopped at frequency ~ 400 Hz and taken to the sample through an oversized hollow circular copper waveguide with a tapered end with the output aperture of 6 mm in diameter fully covering the detector's mesa. The output aperture of this waveguide was located just above the sample so that the incident THz power was distributed fairly homogeneous over the sample area and the value of an energy-flux vector, F, was 3.5 W/m² and paralleled to the normal to the plane of the sample. The electric field of the incident THz wave was polarized across the FET-gate fingers.

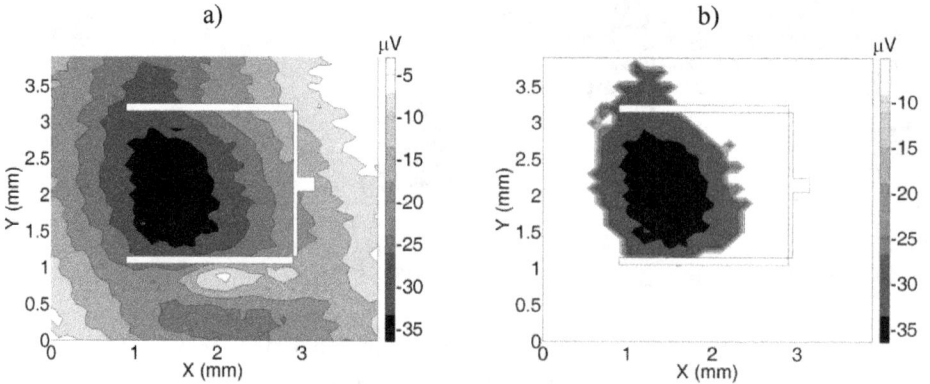

Fig. 2. (a) Raster scan images of photoresponse for grating-gate detector; (b) Effective area image of raster scan image in Fig. 2(a) for grating-gate detector at level $1/\sqrt{2}$. It evidences a small response near common gate metallization.

The source pad was grounded and the gate voltage was applied between the gate and source pads. The photovoltage signal measurement was performed by using a standard lock-in technique.

3. Results and Discussion

3.1. *Mesa resistor in gate circuit*

There are some difference between theoretical calculations and experimental data in Refs. 1 and 7 for grating gate structures. In these papers they were used the theoretical approach developed in Refs. 10 and 11. This method considered only grating gate without included metallization pads that connected fingers of the grating between itself. But in work[12] it has been shown that THz response abruptly drops near a gate bus (Fig. 2 from Ref. 12).

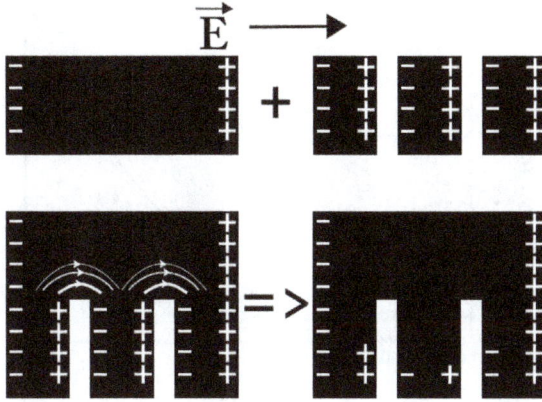

Fig. 3. Schematical picture of the distribution of charges due to incident field for simple metal mirror, grating, for joint mirror+grating, and after joint mirror+grating.

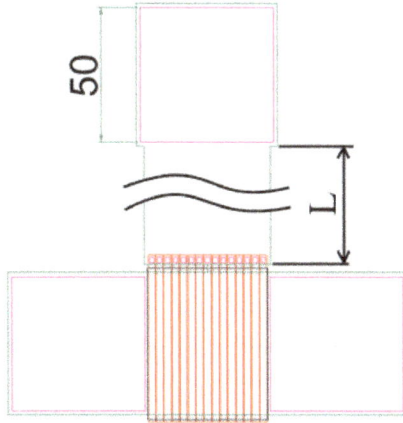

Fig. 4. Top view of the detector (dimension in μm).

The physical explanation is that when a metal bus (mirror in Fig. 3) is attached to the grating, there is flowing current between adjacent fingers of the grating which causes the same potential relief and as for the mirror. So attaching the gate bus to the grating increases the reflectance of the entire structure.

We have made two samples of THz detectors with a grating gate. Fabrication and measurements have been done as in Section 2. Grating gate was made by 150/500 nm Ti-Au layers with 3 μm period and 0.3 μm slit. Top view of the detector is shown in Fig. 4. The detectors have resistors in the circuit of the gate based on mesa to make the metal gate electrode away from the grating. Sample 1 has the length of the resistor $L = 100$ µm, sample 2 – $L = 50$ µm. Measurement of photovoltaic response was carried out at room temperature at a frequency of normally incident radiation 0.6 THz.

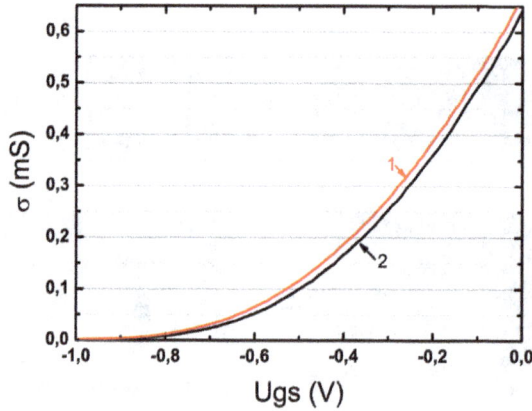

Fig. 5. Transfer characteristics of samples 1 and 2.

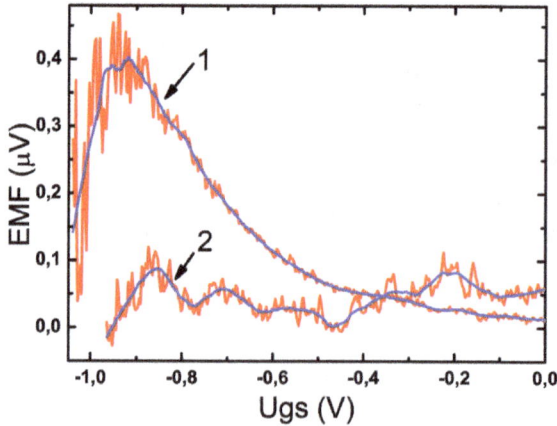

Fig. 6. Experimental graphics of the photoresponse (red) at a frequency of ~ 0.6 THz [averaged photoresponse curves (blue)] for samples 1 and 2.

Figures 5 and 6 show the transfer characteristic of the samples and the response to THz radiation, respectively.

From Fig. 6 it can be seen that for the transistor with a large resistor photoresponse increased, as expected. The influence of bus gate is to reflect of incident THz radiation polarized perpendicular to the grating (and thus parallel to the bus). In studies with a narrow beam of THz radiation the visible effect of reduction of the photoresponse near the contact strips of the grating and the shared bus were exhibited.[12] If THz beam will be covering an entire sample one would be observed the difference in THz photoresponse for large and small resistor sample (Fig. 6). For THz beam directed away the gate bus THz photoresponse will be enhanced.

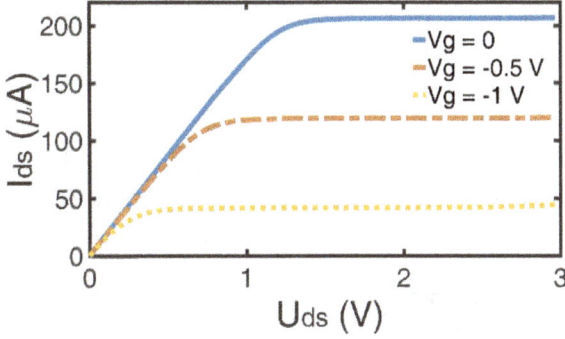

Fig. 7. Voltage-current characteristics of the detector at 4.2 K.

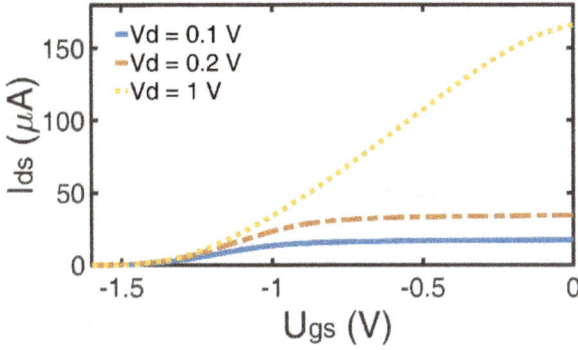

Fig. 8. Transfer characteristics of the detector at 4.2 K.

This effect arises from the fact that the metallization has large conductivity on THz. The skin depth for Au at 0.6 THz is 0.1 μm that gives the resistivity 0.23 Ohm/□. Used resistor in the gate circuit diminishes the current between adjacent fingers because the skin depth for the cap layer at 0.6 THz is 2 nm that gives the resistivity larger than 1 kOhm/□. In the same time, 2DEG has the resistivity 350 Ohm/□ so the gate fingers will be shunted by 2DEG of the resistor. So there is a need to make a long resistor in order to the gate bus does not influence the grating gate. So we have used a mesa resistor to avoid shunting the fingers of the gate electrodes.

3.2. *Results of measurements*

Figures 7 and 8 show the static current-voltage (I-V) and transfer characteristics of the FET chain, respectively.

The channel depletion threshold gate voltage obtained from the interpolation of the linear part of the transfer characteristic down to zero drain current in Fig. 8 is about $U_{th} = -1.3V$.

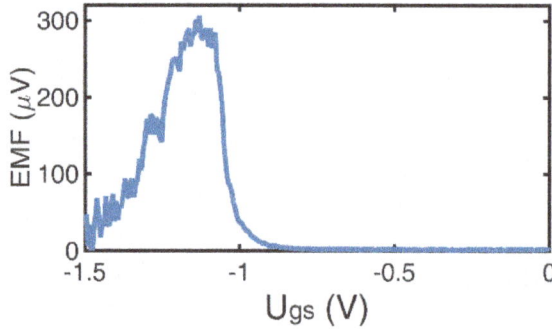

Fig. 9. The experimental graphic of EMF at frequency of ~ 0.6 THz.

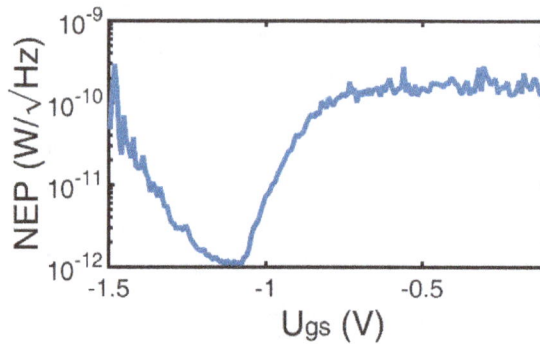

Fig. 10. NEP of the detector.

Measurement of EMF using BWO was carried out at liquid helium temperature at a frequency of normally incident radiation ≈ 0.6 THz (wavelength $\lambda = 500$ μm). Figure 9 shows the response of the detector to THz radiation.

The maximum of EMF is about 300 μV at $U_{gs} \approx -1.15$ V. The behavior of this curve is typical for nonresonant THz response. As far the detector has the asymmetrical structure we have been measured the response without a biased current. The responsivity is calculated as

$$Res = EMF/P,\qquad(1)$$

where *Res* is the responsivity, *EMF* is the electromotive force and *P* is the radiation power incident upon the sample area. Then we have maxima *Res* about 1.4 kV/W. Signal to noise ratio is more important for electro-technical measurement. So we have been evaluated NEP of the detector as

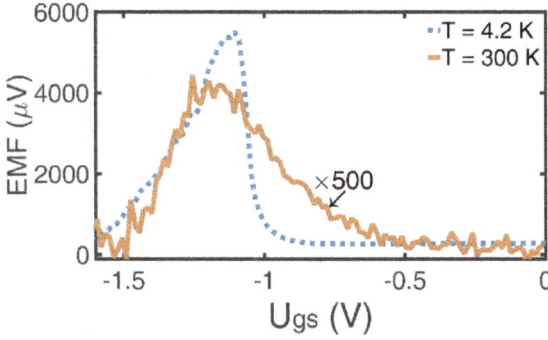

Fig. 11. EMF of the detector at two temperatures (4.2 and 300 K). Room temperature curve is multiplied by 500.

$$NEP = \sqrt{4k_B R_{S-D}}/Res, \tag{2}$$

where k_B is the Boltzmann's constant, R_{S-D} is the source-drain resistance. To calculate *NEP* we used the transfer characteristic at $V_d = 0.1$ V while EMF was measured at $V_d = 0$ so that this value is more correct than the others to use. The dependence *NEP* via U_{gs} is shown on Fig. 10. The minimum of NEP is about 1 pW/$\sqrt{Hz}$ at $U_{gs} \approx -1.1$ V.

The results of our work are similar enough with Ref. 8 what says about the similarity of both types of the detectors. Nevertheless the detector of FET's chain is probably preferable for the magnitude of the NEP and responsivity, especially taking into account the quadratic dependence of the responsivity on frequency in THz range.[13] It should be noted that an improved version of the detector from Ref. 8 with the focusing lens has a better performance of two orders of magnitude in responsivity.[3] The resistance of the channel has a value $\sim 10^5$ Ohm for the gate voltage - 1.1 V. Therefore there is a mismatch of the impedance of free space and the receiver that reduces the responsivity.

To evaluate the response depending on the temperature we used the more powerful radiation source described above in Section 2.2. These measurements are presented in Fig. 11.

It is difficult to compare the response of the present detector with detectors from Refs. 3 and 8 because they have different temperature ranges. But in Ref. 14 it has been shown that a temperature drop of THz response at 40 GHz is more than three orders magnitude for the detector with a lateral dimensions several times smaller than half-wavelength of radiation. Therefore detectors with a size corresponding half-wavelength of radiation can have stable characteristics in a broader temperature range.

4. Conclusion

In conclusion, we fabricated the detector (it is a chip with the two dimensional combination of ohmic and potential contacts) with serial connected single detectors and dimensions corresponding half-wavelength of radiation. We obtained the voltage responsivity above 1.4 kV/W in the unbiased mode of the detector operation, the NEP of the detector is 10^{-12} W/Hz$^{1/2}$ at 0.6 THz and temperature 4.2 K. Such detectors have stable characteristics in a broad temperature range. The used design is not optimal from the point of view of impedance matching of free space and the detector. It needs to match these resistances and the current design has a good potential to do it. Optimal design can allow for further development of such THz detector for a variety of applications.

Acknowledgments

This work was supported by the Russian Federal Target Program "Development of Electronic Component Base and Radio Electronics" (Contract No. 14.427.11.0004), Russian Academy of Sciences Program "Fundamentals and Development of Sensor Nanoelectronics and Devices for Terahertz Frequency Ranges" (No. 0070-2014-0006), and Russian Foundation for Basic Research Program "TeraHertz Sensors for Health Protection" (No. 16-52-76011).

References

1. T. Watanabe, S. Boubanga Tombet, Y. Tanimoto, Y. Wang, H. Minamide, H. Ito, D. Fateev, V. Popov, D. Coquillat, W. Knap, Y. Meziani, and T. Otsuji, "Ultrahigh sensitive plasmonic terahertz detector based on an asymmetric dual-grating gate HEMT structure", *Solid-State Electron.* **78** (2012) 109–114.
2. F. Schuster, D. Coquillat, H. Videlier, M. Sakowicz, F. Teppe, L. Dussopt, B. Giffard, T. Skotnicki, and W. Knap, "Broadband terahertz imaging with highly sensitive silicon CMOS detectors", *Opt. Exp.* **19(8)** (2011) 7827–7832.
3. G. C. Dyer, S. Preu, G. R. Aizin, J. Mikalopas, A. D. Grine, J. L. Reno, J. M. Hensley, N. Q. Vinh, A. C. Gossard, M. S. Sherwin, S. J. Allen, and E. A. Shaner, "Enhanced performance of resonant sub-terahertz detection in a plasmonic cavity", *Appl. Phys. Lett.* **100** (2012) 083506.
4. V. V. Popov, D. M. Yermolaev, K. V. Maremyanin, V. E. Zemlyakov, N. A. Maleev, V. I. Gavrilenko, V. A. Bespalov, V. I. Yegorkin, V. M. Ustinov, and S. Yu. Shapoval, "Detection of terahertz radiation by tightly concatenated InGaAs field-effect transistors integrated on a single chip", *Appl. Phys. Lett.* **104** (2014) 163508.
5. T. A. Elkhatib, V. Y. Kachorovskii, W. J. Stillman, D. B. Veksler, K. N. Salama, X.-C. Zhang, and M. S. Shur, "Enhanced plasma wave detection of terahertz radiation using multiple high electron-mobility transistors connected in series", *IEEE Trans. Microwave Theory Techn.* **58(2)** (2010) 331–339.
6. A. Lisauskas, U. Pfeiffer, E. Öjefors, P. H. Bolìvar, D. Glaab, and H. G. Roskos, "Rational design of high-responsivity detectors of terahertz radiation based on distributed self-mixing in silicon field-effect transistors", *J. Appl. Phys.* **105** (2009) 114511.
7. S. Boubanga-Tombet, Y. Tanimoto, A. Satou, T. Suemitsu, Y. Wang, H. Minamide, H. Ito, D. V. Fateev, V. V. Popov, and T. Otsuji, "Current-driven detection of terahertz radiation using a dual-grating-gate plasmonic detector", *Appl. Phys. Lett.* **104** (2014) 262104.
8. G. C. Dyer, N. Q. Vinh, S. J. Allen, G. R. Aizin, J. Mikalopas, J. L. Reno, and E. A. Shaner, "A terahertz plasmon cavity detector", *Appl. Phys. Lett.* **97** (2010) 193507.
9. D. M. Yermolaev, A. V. Kovalchuk, K. V. Maremyanin, V. I. Gavrilenko, N. A. Maleev, V. M. Ustinov, V. E. Zemlyakov, V. A. Bespalov, V. I. Yegorkin and S. Yu. Shapoval, "Influence of gate bus on terahertz response", *Proc. of 4th Russia-Japan-USA Symp. On Fundamental & Applied Problems of Terahertz Devices & Technologies (RJUS-TeraTech)*, Russia, Chernogolovka, June 2015, pp. 84–85.

10. G. R. Aizin, D. V. Fateev, G. M. Tsymbalov, and V. V. Popov, "Terahertz plasmon photoresponse in a density modulated two-dimensional electron channel of a GaAs/AlGaAs field-effect transistor", *Appl. Phys. Lett.* **91** (2007) 163507.

11. V. V. Popov, "Plasmon excitation and plasmonic detection of terahertz radiation in the grating-gate field-effect-transistor structures", *J. Infrared Millim. Terahertz Waves* **32** (2011) 1178.

12. D. M. Yermolayev, E. A. Polushkin, S. Yu. Shapoval, V. V. Popov, K. V. Marem'yanin, V. I. Gavrilenko, N. A. Maleev, V. M. Ustinov, V. E. Zemlyakov, V. I. Yegorkin, V. A. Bespalov, A. V. Muravjov, S. L. Rumyantsev, and M. S. Shur, "Detection of terahertz radiation by dense arrays of InGaAs transistors", *Int. J. High Speed Electron. Syst.* **24(1&2)** (2015) 1550002.

13. M. Tonouchi, "Cutting-edge terahertz technology", *Nat. Photon.* **1** (2007) 97–105.

14. V. M. Muravev and I. V. Kukushkin, "Plasmonic detector/spectrometer of subterahertz radiation based on two-dimensional electron system with embedded defect", *Appl. Phys. Lett.* **100** (2012) 082102.

Optimization of the Design of Terahertz Detectors Based on Si CMOS and AlGaN/GaN Field-Effect Transistors

Maris Bauer[1], Sebastian Boppel[1,2], Jingshui Zhang[1], Adam Rämer[2], Serguei Chevtchenko[2], Alvydas Lisauskas[1,3,*], Wolfgang Heinrich[2], Viktor Krozer[1,2], and Hartmut G. Roskos[1,†]

[1]*Physikalisches Institut, Goethe-Universität,*
Max-von-Laue-Str. 1, 60438 Frankfurt am Main, Germany

[2]*Ferdinand-Braun-Institut, Leibniz-Institut für Höchstfrequenztechnik,*
Gustav-Kirchhoff-Straße 4, 12489 Berlin, Germany

[3]*Department of Radiophysics, Faculty of Physics,*
Saulėtekio al. 9, bldg. III, Vilnius University, LT-10222 Vilnius, Lithuania
[*]lisauskas@physik.uni-frankfurt.de
[†]roskos@physik.uni-frankfurt.de

TeraFETs are THz power detectors based on field-effect transistors (FETs) integrated with antennas. The first part of this paper discusses the design of Si CMOS TeraFETs leading to an optimized noise-equivalent power close to the room-temperature limit. The impact of the choice of the gate width and gate length, the role of the parasitic effects associated with the technology node, and the conjugate matching of antenna and FET impedance – which is possible over narrow THz frequency bands because of the frequency dependence of the channel impedance resulting from plasmonic effects – are highlighted. Taking these aspects into account, we implement narrow-band detectors of two different designs. Using a 90-nm and a 65-nm CMOS technology, we reach a room-temperature cross-sectional NEP of 10 pW/√Hz at 0.63 THz. We then explore the optimization of AlGaN/GaN TeraFETs equipped with broadband antennae. A room-temperature optical NEP of 26 pW/√Hz is achieved around 0.5 THz despite the fact that the existence of pronounced ungated regions leads to a significant hot-electron thermoelectric DC voltage reducing the rectified signal. AlGaN/GaN TeraFETs become competitive and they have the added advantage that they are extraordinarily robust against electrostatic shock even without inclusion of protection diodes into the design.

Keywords: Terahertz; field-effect transistors; distributed resistive and plasmonic mixing; noise-equivalent power; hot-electron thermoelectric effect.

1. Introduction

TeraFET power detectors operate on the basis of rectification of THz waves coupled into the channel of a field-effect transistor. The mechanism at low frequencies is resistive mixing, but with rising frequency, charge-density waves in the channel play an increasingly important role.[1-3] An aspect whose significance cannot be overestimated is that highly sensitive TeraFETs can be fabricated conveniently by Si CMOS foundry technologies. This has accelerated developments, which have already led to the exploration of camera[4-7] and sensor[8] applications, while detector optimization is still ongoing, gradually leading up to fundamental performance limits for room temperature operation of the devices.[7,9] While Si CMOS TeraFETs are yet unsurpassed in noise-equivalent power (NEP), a number of other materials and technologies are being explored. An example is the AlGaN/GaN HEMT technology,[10-12] which allows to fabricate TeraFETs with a high robustness against electrostatic shock and high THz field amplitudes.

The next section describes the basic design rationale for the achievement of optimal sensitivity and NEP performance of antenna-coupled TeraFETs.

2. TeraFET Design Considerations

For defined absorption of the free-space radiation, the FETs are connected to an integrated antenna. We discuss the optimal choice for the fundamental transistor geometry parameters channel length, channel width, as well as the choice of technology generation particularly for Si CMOS, and point out the possibility of antenna impedance matching.

(i) Gate length: When FETs are used as THz detectors above their transit-time-determined cutoff frequency, charge density (plasma) waves are launched from one side of the gated channel region. These waves usually have a decay length L_D of a few tens of nm in the terahertz range (e.g., L_D = 42 nm at 600 GHz).[3] In case of channel lengths longer than L_D, the waves die out before reaching the other side of the channel. This detection regime is called *non-resonant mixing regime.* The rectification happens only at the beginning of the channel, where the wave still exists. Therefore, the wave's voltage and current amplitudes – along with the transistor's ac impedance – and the rectification process are not much affected by shortening of the channel as long as it remains longer than L_D. If the channel length is reduced to L_D or even made shorter, then the mixing efficiency diminishes. As long as the second port of the channel (the one where the THz signal is not coupled in) is ac-shunted, the mixing efficiency remains equal to that of resistive mixing and only the plasmonic enhancement is missing.[13] If the port is not ac-shorted, then the signal deteriorates more. Hence, best performance is expected for $L \geq L_D$.[a]

The achievable electrical current responsivity can be estimated as follows. In the quasi-static and deep-sub-threshold limits, the current responsivity can be shown to be have an ideal value of $1/2\eta V_T \approx 20/\eta$ [A/W],[13] with η being the ideality factor (a typical value of which is $\eta = 1.77$)[15] and V_T the threshold voltage. This responsivity is identical to that of diode-based rectifiers. Above the cut-off frequency, this limit scales with the ratio of the ac to the dc impedance of the transistor's channel. Since the ac impedance does not change with channel length, and the dc impedance is proportional to it, the current responsivity scales with the inverse of the channel length.

---

[a]Note that the resonant detection regime,[1] where the mean-free path of the plasma wave is much longer than L_D, and the channel may become a resonant cavity for the plasma waves, is not discussed here. With the material systems studied until now, this regime has only relevance at cryogenic temperatures.[14]

The NEP is the ratio of the detector's noise spectral density and its current (or voltage) responsivity. The dominating contribution to the current noise in TeraFETs operated with zero drain bias is Johnson-Nyquist noise which is inversely proportional to the square root of the channel length. With the current responsivity's inverse dependence on channel length, the NEP at frequencies well above the cut-off frequency is being reduced and thus becomes better if the channel length is lowered.

(ii) Gate width: The electrical current responsivity is independent of the transistor's channel width W, because the ac and dc impedance both inversely scale with W. Since the thermal current noise is proportional to the square root of the width, it can be concluded that the smallest possible transistor width should be chosen for maximum responsivity. However, with decreasing width, the transistor's ac impedance becomes too large, and it will be impossible to impedance-match the antenna to the channel, i.e., only a limited amount of power can be delivered from the antenna to the transistor. For this reason, a wider channel can be advisable at lower terahertz frequencies (discussion below).

(iii) CMOS technology generation: We briefly discuss how much performance improvement can be expected by the choice of an advanced Si CMOS technology. Generally, TeraFET development benefits from advances in the semiconductor industry due to a constant reduction of parasitic resistance and capacitance effects, which are limiting factors of high-speed device operation. Table 1 illustrates the technology development with data of the Predictive Technology Model (PTM) published in 2006.[16] The technological nodes addressed there have been used for the fabrication of most of the Si CMOS TeraFETs reported about in the literature.

Table 1. Parameters for different Si CMOS technology nodes. The parameters are from the Predictive Technology Model (PTM).[16] L_{min} denotes the minimal gate lengths, T_{ox} the thickness of the gate insulator, R_{sdw} the specific resistance of the source-drain parasitic resistance, C_{par} the sum of junction, overlap and fringe capacitance per unit length, and μ the channel mobility.

L_{min}	T_{ox}	R_{sdw}	C_{par}	μ
nm	nm	$\Omega\mu m$	fF/μm	cm²/Vs
130	2.25	180	1	300
90	2.05	170	1	270
65	1.85	160	1	240
45	1.75	155	1	225
32	1.65	150	1	210

As stated above, the current responsivity is independent of the gate width W, while it scales inversely with the channel length. For the latter aspect, it improves roughly proportionally with the inverse of L_{min}, the minimal gate length of the respective technology node. In addition to the current responsivity, thermal noise is considered by

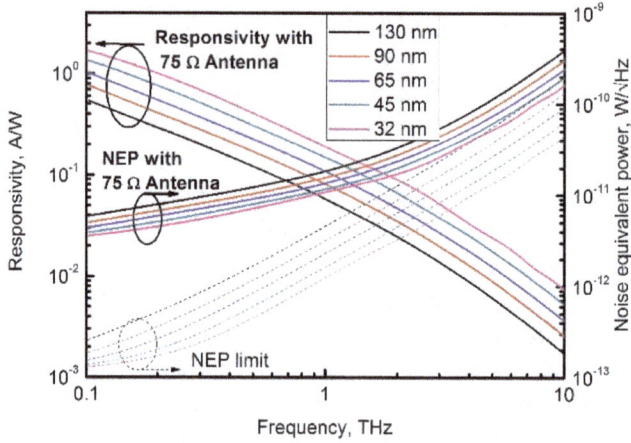

Fig. 1. Calculated cross-sectional responsivity and NEP versus frequency for five technology nodes (130 nm down to 32 nm). Solid lines show the responsivity and NEP if the FET is connected to a 75-Ω lossless broadband antenna. The gate width W is optimized at each frequency for minimal NEP. For comparison, the theoretical NEP limits are shown as dashed lines assuming impedance matching to a transistor with the shortest gate length allowed by the respective technology and a gate width set to be $W = 1.5\ L$.

calculating the theoretically achievable cross-sectional NEP[b] limit for each parameter set of the respective technological node. Figure 1 displays these minimal cross-sectional NEPs as dashed lines. The calculations assume (i) the gate voltage to be the threshold voltage, (ii) the power source to be impedance-matched to the transistor's input impedance, (iii) the channel length to be the smallest possible for each node, and (iv) the gate width set to be $W = 1.5\ L$. At 100 GHz, the theoretical limit of the NEP for the 32-nm technology is found to be 0.1 pW/$\sqrt{Hz}$, increasing to 10 pW/$\sqrt{Hz}$ at 4 THz. For the 130-nm node, these values rise to 0.25 pW/$\sqrt{Hz}$ and 40 pW/$\sqrt{Hz}$, respectively. At 1 THz, NEP values in the range of 1-6 pW/$\sqrt{Hz}$ are feasible.

(iv) Antenna and impedance matching for Si CMOS TeraFETs: In practice, for Si CMOS, the discussed theoretical limits cannot be reached at lower THz frequencies, since the ac device impedance becomes too large for conjugate matching with realistic antennas. The low-frequency channel resistance of the smallest possible device in each technology generation is on the order of a few ten kΩ, monotonously decreasing with rising frequency above the transistor's cutoff frequency due to the rising influence of the charge-density waves in the channel.[13] Typically, slightly below 1 THz, the channel's ac

---

[b]The *cross-sectional* responsivity and NEP take only that part of the incident power into account which impinges within the cross-section of the antenna (which is either measured or derived from numerical simulations). In comparison, the full incident power enters into the determination of the *optical* responsivity and NEP. The former quantity is often considered if the design of the detector prevents the use of a substrate lens. In this paper, the Si CMOS detectors have antennae with buried ground planes which make a THz illumination through the substrate impossible. The THz radiation is hence coupled in from the top. As the top side holds the contact wires, a substrate lens cannot be attached readily. In the case of the AlGaN/GaN devices, bow-tie or log-spiral antennae are implemented, thus illumination through the substrate and an attached Si substrate lens is employed.

resistance drops below 1 kΩ and impedance matching becomes feasible. For discrete sub-THz design frequencies and narrow-band antennae with radiation resistances of several hundred Ω, it is often possible to reach impedance matching by proper choice of the width of the transistor. In general, better performance at any given target frequency can be achieved with high impedance antennae, at the cost of narrower impedance bandwidths.

Common broadband antennas, such as log-spiral, log-periodic and bow-tie antennas, exhibit an antenna impedance on the order of 75 Ω. Hence, for a reasonable comparison of the achievable NEPs of the different technological nodes, we consider a lossless 75-Ω antenna and optimize the gate width for the lowest NEP value per frequency. Corresponding results are shown as full lines in Fig. 1. One finds that, at low frequencies, wider transistors are favorable, although at high frequencies, when the transistor impedance is dominated by drain and source parasitic resistances, narrow-width transistors close to $W = 1.5\,L$ are optimal. Taking these considerations into account, we see only a moderate performance gain with regard to the NEP when more advanced technology nodes are used. This is illustrated in Fig. 1, where the NEP value rises by not more than a factor of 2.5 if the technology node is changed from 130 nm to 32 nm.

3. Si CMOS TeraFET Implementation

Following the above design rationale, and aiming for conjugate impedance matching of antenna and FET, a resonant patch-antenna-based detector was realized in 90-nm CMOS technology reaching a cross-sectional voltage responsivity of up to 2.55 kV/W and a minimal cross-sectional NEP value of 10 pW/√Hz (calculated 8.5 pW/√Hz) at 630 GHz. The antenna impedance is 220 Ω (simulated). A second design employs a λ-dipole with ground plane. This design was implemented in a 65-nm CMOS technology and leads to an even higher antenna resistance between 600 and 700 Ω (simulated). The measured minimal NEP is 13.6 pW/√Hz, the calculated one 8.8 pW/√Hz. The measured maximal responsivity is 2.8 kV/W. More details will be published elsewhere.

4. Broadband AlGaN/GaN HEMT Detectors

In addition to CMOS TeraFETs, we also investigate broadband TeraFET detectors in AlGaN/GaN HEMT technology with integrated bow-tie or log-spiral antennae. One advantage of these detectors is that they are extremely robust against external electrostatic influences and intense transient fields. Again, impedance matching over a wide band of frequencies is impossible to achieve, and for that reason, the overall performance of broadband TeraFETs tends to be inferior to detectors with resonant antennae.

Optimized AlGaN/GaN TeraFETs were designed to span the frequency range from 0.5 to 1.18 THz. Extensive studies were made with a technology run where the transistors had a gate length of 250 nm and a total source-drain distance of 950 nm (bow-tie antenna) respectively 850 nm (log-spiral antenna). The channel width was 3 μm. Table 2 lists measured detector sensitivities over the design frequency range. We reached high sensitivities, i.e., low optical NEP, down to 57 pW/√Hz at 0.9 THz for the bow-tie design and down to 42 pW/√Hz at 0.7 THz for the log-spiral design,[11] noting that the latter couples only ~ 50 % of linearly polarized THz radiation, which has not been corrected for in Table 2.

Table 2. Measured AlGaN/GaN TeraFET optical NEPs in pW/√Hz.

Freq. [THz]	0.5	0.6	0.7	0.8	0.9	1	1.08	1.18
Bow-tie	82	79	83	63	57	71	90	155
Log-Spiral	98	68	43	56	65	85	93	169

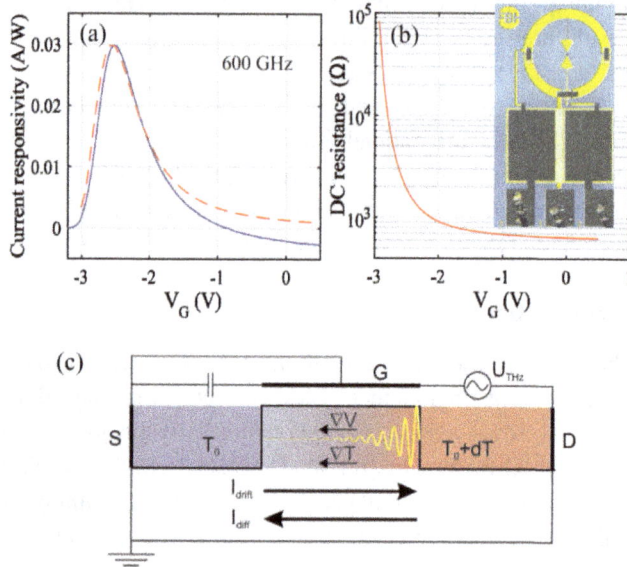

Fig. 2. (a) Optical current responsivity of an AlGaN/GaN TeraFET with integrated bow-tie antenna at 600 GHz (blue full line) and anticipated response calculated from the dc resistance (red dashed line). (b) Dc drain-source resistance, inset: microscopic image of the detector. (c) Schematic representation of the detector showing the THz coupling scheme and induced detection signal contributions. The yellow wavy line in the gated region represents the plasma wave.

From the optical NEP values, we can determine the electrical NEP of the devices by correction for the antenna coupling losses and the antenna impedance mismatch. The combined correction factor is estimated to be roughly 2 for the bow-tie design. The calculated electrical NEP is 27.4 pW/√Hz at 0.9 THz. For further details about the antenna design, the detector characterization, and the calculation of the device NEP, see Ref. 11.

In a very recent technology run, bow-tie detector designs with transistors with a shorter gated region (100 nm instead of 250 nm) and ungated regions of 300 nm on each side were tested. For a series of devices, we measured an average optical NEP of 40 pW/√Hz at 0.5 THz and optimal gate biasing, with the best optical NEP value of one device being found to be 26 pW/√Hz.

The measured current responsivity versus gate bias voltage for a detector with integrated bow-tie antenna at 600 GHz of Ref. 11 is shown in Fig. 2(a) together with the dc drain-source resistance in Fig. 2(b). Figure 2(c) displays a photographic and schematic representation of a bow-tie TeraFET. For efficient detection, it is essential that the induced plasma waves are launched only from one side into the gated transistor region. Otherwise, rectification takes place on both ends of the channel, and the signals cancel due to opposite signs of the induced rectified currents and voltages. In the case of our AlGaN/GaN HEMT devices, capacitive shunting of the source and gate contacts by a MIM capacitance under one of the antenna arms ensures proper boundary conditions at all frequencies.[11,12]

It is known that the knowledge of the dc drain-source resistance allows to make a fairly good prediction of the THz responsivity of TeraFETs in the distributed resistive mixing regime using an expression derived in Ref. 17. The corresponding calculated THz response of the bow-tie detector is plotted in Fig. 2(a) as a dashed line (normalized to the experimental curve at the point of best NEP). At high gate voltage, it exhibits a significant deviation from the measured signal, which manifest itself in two main aspects, (i) a zero-crossing of the detection signal, and (ii) that the signal does not converge to zero for rising gate voltage. These features exhibit a pronounced frequency dependence (not shown here).

Detailed analysis supports a hot-electron thermoelectric origin of these features. Figure 2(c) illustrates its origin. As the source side of the transistor channel is pinned to ground at THz frequencies by the MIM capacitor, only the electrons on the drain side of the transistor are locally heated by the incoming THz radiation. This gives rise to a gradient of the charge-carrier temperature along the channel whose exact spatial dependence results from the distribution of the absorbed THz power along the ungated drain region and the channel region. As a result, a thermal diffusion current with opposite sign with respect to the plasma wave-induced drift current is generated, which counteracts the classical Dyakonov-Shur mixing and produces the observed measurement characteristics. First simulations with an extended drift-diffusion model indicate that the main contribution to the thermoelectric signal is due to the temperature difference of the ungated regions, indicated as T_0 and T_0+dT in Fig. 2(c), the difference being determined by the impedance (power absorption) conditions of the various regions.

We observe this additional thermoelectric signal contribution with TeraFETs implemented in various material systems, such as the presented AlGaN/GaN HEMTs, graphene FETs[18] as well as carbon-nanotube FETs,[19] all operated at room temperature (data not shown). For Si CMOS TeraFETs, we find thermoelectric signatures only for cryogenic conditions and not at room temperature. This different behavior of CMOS TeraFETs may be related to the self-alignment technology used for the fabrication of the source and drain contacts which leads to virtually non-existent ungated regions in MOSFETs.

5. Conclusion

We have discussed design aspects for the implementation of antenna-coupled field-effect transistors for THz detection. Besides investigating the importance of the device geometry parameters, we have also evaluated fundamental constraints of available Si CMOS technologies and their influence on the performance of the detectors. In general, smaller transistors (in width and length) are favorable. However, depending on the design frequencies and the particular integrated antenna, narrow transistors can exhibit such high impedances that impedance matching and therefore power delivery to the transistor channel might become unfavorable or impossible at low sub-THz frequencies. We find that it is not mainly the particular technology node that limits the achievable performance but the feasibility of impedance matching of the integrated antenna to the transistor channel, which can favor less-advanced technologies with a higher degree of freedom with regard to the choice of the design parameters. Based on these considerations, we have realized two types of high-sensitive TeraFET detectors, i.e., a resonant patch-antenna-coupled detector in Si CMOS and broadband antenna-coupled devices with AlGaN/GaN HEMTs. With the resonant design, we achieve a maximum cross-sectional NEP of 10 pW/$\sqrt{\text{Hz}}$ at 0.63 THz. With the broadband designs, we find minimum optical NEPs of 26 pW/$\sqrt{\text{Hz}}$ at 0.5 THz for a bow-tie design, all for linearly polarized radiation. To the authors' knowledge, these values are record values for TeraFETs in GaN technology and among the best for Si technology. While not yet reaching the NEP performance of Si CMOS, AlGaN/GaN TeraFETs are clearly becoming competitive. Notably, their handling in laboratory environments is much more convenient because of their insensitivity to electrostatic influences.

We have addressed experimental evidence that thermoelectric diffusion currents – in addition to (charge-density-wave-enhanced) resistive rectification – contribute to the THz detector signal. Since we observe these contributions in a number of material/technology systems (albeit for Si CMOS TeraFETs only at cryogenic temperatures), diffusive currents should be included in the theoretical description of rectification in TeraFETs. This constitutes an extension to present modeling approaches, which neglect these terms in the hydrodynamic transport equations.

Acknowledgments

This work was financially supported by the Leibniz-Gemeinschaft, the German-Israeli Foundation for Scientific Research and Development, the Hessian Excellence Initiative LOEWE (projects: *Sensors towards Terahertz* and *Terahertz-Kamera für die zivile Sicherheitstechnik*), and the EU project HYPERIAS. J.Z. acknowledges a Ph.D. scholarship of the Chinese Scholarship Council.

References

1. M. Dyakonov and M. Shur, "Detection, mixing, and frequency multiplication of terahertz radiation by two-dimensional electronic fluid", *IEEE Trans. Electron. Dev.* **43** (1996) 380–387.

2. R. Tauk, F. Teppe, S. Boubanga, D. Coquillat, W. Knap, Y. M. Meziani, C. Gallon, F. Boeuf, T. Skotnicki, C. Fenouillet-Beranger, D. K. Maude, S. Rumyantsev, and M. S. Shur, "Plasma wave detection of terahertz radiation by silicon field effects transistors: Responsivity and noise equivalent power", *Appl. Phys. Lett.* **89** (2006) 253511.
3. A. Lisauskas, U. Pfeiffer, E. Öjefors, P. Haring Bolívar, D. Glaab, and H. G. Roskos, "Rational design of high-responsivity detectors of terahertz radiation based on distributed self-mixing in silicon field-effect transistors", *J. Appl. Phys.* **105** (2009) 114511.
4. R. Al Hadi, H. Sherry, J. Grzyb, Y. Zhao, W. Förster, H. M. Keller, A. Cathelin, A. Kaiser, and U. R. Pfeiffer, "A 1 k-pixel video camera for 0.7-1.1 terahertz imaging applications in 65-nm CMOS", *IEEE J. Solid-State Circ.* **47** (2012) 2999–3012.
5. J. Zdanevičius, M. Bauer, S. Boppel, V. Palenskis, A. Lisauskas, V. Krozer, and H. G. Roskos, "Camera for high-speed THz imaging", *J. Infrared Milli. Terahertz Waves* **36** (2015) 986-997.
6. S. Boppel, A. Lisauskas, and H. G. Roskos, "Terahertz array imagers: towards the implementation of terahertz cameras with plasma-wave-based silicon MOSFET detectors", in *Handbook of Terahertz Technology for Imaging, Sensing, and Communications*, ed. D. Saeedkia, Woodhead Publishing Ltd., 2013, pp. 231-271.
7. D. Y. Kim, S. Park, R. Han, and K. K. O, "Design and demonstration of 820-GHz array using diode-connected NMOS transistors in 130-nm CMOS for active imaging", *IEEE Trans. THz Sci. Technol.* **6** (2016) 306–317.
8. M. Bauer, R. Venckevičius, I. Kašalynas, S. Boppel, M. Mundt, L. Minkevičius, A. Lisauskas, G. Valušis, V. Krozer, and H. G. Roskos, "Antenna-coupled field-effect transistors for multi-spectral terahertz imaging up to 4.25 THz", *Opt. Exp.* **22** (2014) 19250–19256.
9. U. R. Pfeiffer, J. Grzyb, H. Sherry, A. Cathelin and A. Kaiser, "Toward low-NEP room-temperature THz MOSFET direct detectors in CMOS technology", in *Proc. 38th IRMMW-THz*, Mainz, Germany, Sep. 2013.
10. J. D. Sun, Y. F. Sun, D. M. Wu, Y. Cai, H. Qin, and B. S. Zhang, "High-responsivity, low-noise, room-temperature, self-mixing terahertz detector realized using floating antennas on a GaN-based field-effect transistor", *Appl. Phys. Lett.* **100** (2012) 013506.
11. M. Bauer, A. Rämer, S. Boppel, S. Chevtchenko, A. Lisauskas, W. Heinrich, V. Krozer, and H. G. Roskos, "High-sensitivity wideband THz detectors based on GaN HEMTs with integrated bow-tie antennas", in *Proc. 10th Europ. Microw. Integrated Circuits Conf. (EuMIC)*, Paris, France, Sep. 2015.
12. S. Boppel, M. Ragauskas, A. Hajo, M. Bauer, A. Lisauskas, S. Chevtchenko, A. Rämer, I. Kašalynas, G. Valušis, H.-J. Würfl, W. Heinrich, G. Tränkle, V. Krozer, and H. G. Roskos, "0.25-μm GaN TeraFETs optimized as THz power detectors and intensity-gradient sensors", *IEEE Trans. THz Sci. Technol.* **6** (2016) 348–350.
13. S. Boppel, A. Lisauskas, M. Mundt, D. Seliuta, L. Minkevičius, I. Kašalynas, G. Valušis, M. Mittendorf, S. Winnerl, V. Krozer, and H. G. Roskos, "CMOS integrated antenna-coupled field-effect transistors for the detection of radiation from 0.2 to 4.3 THz", *IEEE Trans. Microwave Theory Techn.* **60** (2012) 3834–3843.
14. W. Knap, M. Dyakonov, D. Coquillat, F. Teppe, N. Dyakonova, J. Łusakowski, K. Karpierz, M. Sakowicz, G. Valušis, D. Seliuta, I. Kašalynas, A. El Fatimy, Y. M. Meziani, and T. Otsuji, "Field effect transistors for terahertz detection: Physics and first imaging applications", *J. Infrared Milli. Terahz Waves* **30** (2009) 1319–1337.
15. A. Lisauskas, S. Boppel, J. Matukas, V. Palenskis, L. Minkevičius, G. Valušis, P. P. Haring Bolívar, and H. G. Roskos, "Terahertz responsivity and low-frequency noise in biased silicon field-effect transistors", *Appl. Phys. Lett.* **102** (2013) 153505.
16. W. Zhao and Y. Cao, "New generation of predictive technology model for sub-45 nm early design exploration", *IEEE Trans. Electron. Dev.* **53** (2006) 2816–2823, and *Proc. 7th Int. Symp. on Quality Electronic Design (ISQED)*, San Jose, USA, March 2006, p. 585.

17. M. Sakowicz, M. B. Lifshits, O. A. Klimenko, F. Schuster, D. Coquillat, F. Teppe, and W. Knap, "Terahertz responsivity of field effect transistors versus their static channel conductivity and loading effects", *J. Appl. Phys.* **110** (2011) 054512.
18. A. Zak, M. A. Andersson, M. Bauer, J. Matukas, A. Lisauskas, H. G. Roskos, and J. Stake, "Antenna-integrated 0.6 THz FET direct detectors based on CVD graphene", *Nano Lett.* **14** (2014) 5834–5838.
19. M. Bauer, A. Lisauskas, P. Sakalas, M. Schröter, and H. G. Roskos, "Terahertz detection at 240 GHz with a semiconducting carbon-nanotube field-effect transistor", in *Proc. 39th IRMMW-THz*, Tucson, Arizona, USA, Sep. 2014.

Design and Fabrication of Terahertz Detectors Based on 180-nm CMOS Process Technology

Kosuke Wakita and Eiichi Sano

Research Center for Integrated Quantum Electronics, Hokkaido University,
North13, West8, Sapporo, Hokkaido, 060-0814, Japan
{wakita, esano}@rciqe.hokudai.ac.jp

Masayuki Ikebe

Graduate School of Information Science and Technology, Hokkaido University,
North14, West9, Sapporo, Hokkaido, 060-0814, Japan
ikebe@ist.hokudai.ac.jp

Stevanus Arnold and Taiichi Otsuji

Research Institute of Electrical Communication, Tohoku University,
2-1-1 Katahira, Aoba, Sendai, Miyagi, 980-8577, Japan
{arnold, otsuji}@riec.tohoku.ac.jp

Yuma Takida and Hiroaki Minamide

Tera-Photonics Research Team, RIKEN Center for Advanced Photonics, RIKEN,
519-1399 Aramaki-aza Aoba, Sendai, Miyagi, 980-0845, Japan
{yuma.takida, minamide}@riken.jp

A CMOS cascode amplifier, biased near the threshold voltage of a MOSFET, for terahertz direct detection is proposed. A CMOS terahertz imaging circuit (size: 250 x 180 μm) is designed and fabricated on the basis of low-cost 180-nm CMOS process technology. The imaging circuit consists of a microstrip patch antenna, an impedance-matching circuit, and a direct detector. It achieves a responsivity of 51.9 kV/W at 0.915 THz and a noise equivalent power (NEP) of 358 pW/Hz$^{1/2}$ at a modulation frequency of 31 Hz. NEP is estimated to be reduced to 42 pW/Hz$^{1/2}$ at 100 kHz. These results suggest that cost-efficient terahertz imaging is possible in the near future.

Keywords: CMOS; detector; imaging; responsivity; NEP; terahertz; THz; on-chip patch antenna; microstrip antenna; direct detection; envelope detection.

1. Introduction

The terahertz-frequency region (100 GHz – 10 THz), namely, between millimeter-waves and far-infrared light waves, has attracted much attention owing to its wide range of applications (such as wireless communications and sensors). Terahertz waves, which pass through a wide variety of substances, such as plastics, fabrics, and papers, can reveal hidden contents that cannot be seen by visible light.[1-3] In addition, so-called fingerprint spectra in the terahertz frequencies can identify hazardous materials. However, the lack of low-cost and small-size microelectronics that generate sufficient power and detect a faint signal (often called the "THz gap") is one of the major obstacles preventing terahertz applications from coming into wide use in our daily life.[4] In the microelectronics community, terahertz detection outperforms terahertz generation. Detection methods are generally divided into two kinds, namely, coherent (heterodyne) detection and incoherent (direct) detection. Although heterodyne detection achieves higher sensitivity than that for direct detection, it requires a set of a local oscillator and a mixer that is difficult to construct with today's technology, even if a sub-harmonic scheme is used. For that reason, direct detection has been used. Terahertz direct detectors have primarily relied on specialized fabrication technologies such as Schottky diodes,[5] bolometers,[6] and high-electron-mobility transistors.[7] Many of these technologies require additional process steps to make them compatible with CMOS technologies.[8] Silicon-CMOS process technologies are becoming a cost-efficient alternative. The main advantage of silicon technologies is that they allow low-cost and large-scale integration of circuits by using readout electronics and on-chip signal processors.[9] Furthermore, the high-frequency capabilities of silicon technologies have steadily been improved in accordance with the guiding principle of scaling, which enables high integration level and low power consumption at higher frequencies.[10]

In this study, a CMOS cascode amplifier biased near the threshold voltage of a MOSFET is used to increase the responsivity of terahertz direct detectors. The detection principle is explained in Section 2, a CMOS terahertz imaging circuit is described in Section 3, measurement results are presented in Section 4, and a summary and conclusion are given in Section 5.

2. Detection Principle

For terahertz-wave detection, non-linearity of an n-MOSFET is utilized. The drain current of the n-MOSFET, which is biased in the saturation region, is given by

$$I_{DS} = \frac{\mu_0 C_{OX} W}{2L} (V_{GS} - V_T)^2,$$ (1)

where W and L are width and length of the channel, C_{OX} is capacitance per unit area of the gate oxide, μ_0 is surface mobility of the channel, V_{GS} is gate voltage, and V_T is threshold voltage.[11] Gate voltage is expressed by

$$V_{GS} = V_0 + v \sin \omega t,$$ (2)

where V_0 is input bias voltage, and $v \sin \omega t$ is input small signal. Combining Eqs. (1) and (2) gives

$$I_{DS} = \frac{\beta}{2} \left\{ (V_0 - V_T)^2 + v^2 \sin^2 \omega t + 2(V_0 - V_T)v \sin \omega t \right\},$$ (3)

where transconductance parameter β is given by

$$\beta = \frac{\mu_o C_{OX} W}{L} \,. \tag{4}$$

Cutoff frequency of a MOSFET in the case of the 180-nm CMOS technology is much lower than the terahertz-frequency region. Therefore, a time-harmonic AC input signal with a period of T is averaged as

$$\lim_{T \to \infty} \frac{\int_0^T \sin^2 \omega t \, dt}{T} = \frac{1}{2}, \quad \lim_{T \to \infty} \frac{\int_0^T \sin \omega t \, dt}{T} = 0 \,. \tag{5}$$

When a terahertz signal is input, the response of the drain current is given by

$$I_{DS} = \frac{\beta}{2} \left\{ (V_0 - V_T)^2 + \frac{v^2}{2} \right\}. \tag{6}$$

In Eq. (6), the amplitude component of the signal remains. Therefore, the n-MOSFET acts as an envelope detector for carrier frequencies exceeding the cutoff frequency of the MOSFET. The detection principle is explained schematically in Fig. 1.

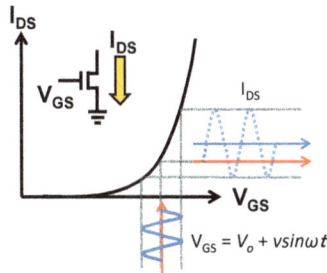

Fig. 1. Non-linear detection scheme.

3. Design of Imaging Circuit

A block diagram of the imaging circuit composed of a microstrip patch antenna, matching circuit, and detector is shown in Fig. 2.

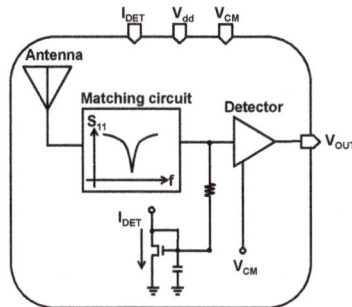

Fig. 2. Block diagram of the imaging circuit.

3.1. *Detector*

Three types of terahertz detectors were designed and fabricated. The circuits shown in Fig. 3 were used as envelope detectors. Type A is a cascode amplifier. Type B consists of a fundamental amplifier and a subthreshold-biased operational amplifier (subVth-OP amp), while in Type C, the fundamental amplifier is replaced by the cascode amplifier. The cascode amplifier consists of a common-source n-MOSFET (as the input stage) driven by terahertz signal V_{IN} with a common-gate n-MOSFET. The p-MOSFET M1 in each type is the load. The cascode-amplifier circuit operates as an envelope detector, as explained in Section II. The subVth-OP amp operates as a feedback circuit, which determines the load resistance of M1.[12] Output voltage of the detector is fixed to common-mode voltage V_{CM} at the DC bias. Since the subVth-OP amp operates very slowly with a large time constant, the feedback operation is established only at DC and very low frequencies. The feedback circuit in the detector operates as a high-pass filter. The envelope detector with the feedback subVth-OP amp therefore produces no DC offset voltage as its output.

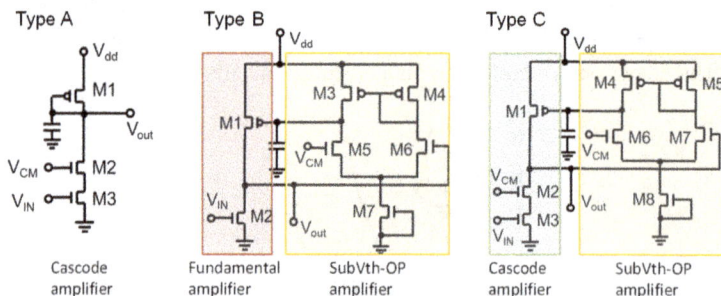

Fig. 3. Schematic of detector circuit.

Measured output noise and gain of three types of the detectors are shown in Fig. 4. The output noise from the detector was amplified by an external low-noise amplifier (NF Corporation, SA-421F5) and analyzed with a spectrum analyzer (Keysight Technologies, E4448A). The output gain was measured with a frequency-response analyzer (NF Corporation, FRA5097).

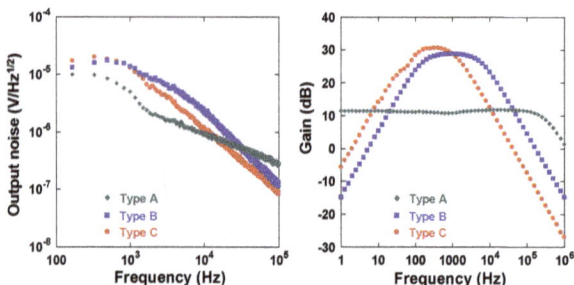

Fig. 4. Measured output noise and gain.

The output noise of Type A is the smallest at low frequency, while that of Type C is the smallest at high frequency. Types B and C have gain roll-off characteristics below 100 Hz due to the feedback operation of the subVth-OP amp, and they produce no DC offset voltage as their outputs. The simulated responsivity of type C is one-order larger than those of types A and B. Accordingly from the viewpoint of noise and responsivity, Type C detector was chosen for the terahertz imaging circuit.

3.2. *Microstrip patch antenna and matching circuit*

Each pixel was equipped with a microstrip patch antenna formed by a metal-6 layer (top aluminum layer) and a ground plane (metal-1 layer) based on the 180-nm CMOS technology. The dimensions of the antenna were determined with a commercial electromagnetic simulator (Keysight Technologies, EMPro) to operate at around 0.9 THz.

A microstrip line was used for the impedance-matching circuit because it is difficult to use spiral inductors and metal-insulator-metal capacitors in the case of terahertz region.

4. Measurement Results

A die photo of the fabricated imaging circuit is shown in Fig. 5. It consists of a microstrip patch antenna, a matching circuit, and type C detector. Its total size is 250 x 180 μm^2.

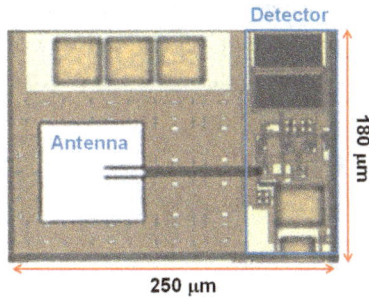

Fig. 5. Die photo of fabricated imaging circuit.

The setup for measuring the responsivity of the fabricated imaging circuit is shown in Fig. 6. An injection-seeded terahertz-wave parametric generator (is-TPG) was used as a frequency-tunable terahertz-wave source[13]. By changing the noncollinear phase-matching condition in the MgO:LiNbO$_3$ crystal, the terahertz-wave frequency was continuously tuned from 0.8 to 1.1 THz with a linewidth of approximately 4 GHz. At 0.905 THz, the terahertz-wave pulse energy was measured to be 132 nJ/pulse with a repetition rate of 100 Hz, corresponding to an average power of 13.2 μW. According to the measured terahertz-wave beam profile, the average power irradiated to the microstrip patch antenna was estimated to be 4.96 nW. Terahertz-wave output was modulated at 31 Hz by an optical chopper, and lock-in detection with a 10-100 Hz bandpass filter was used.

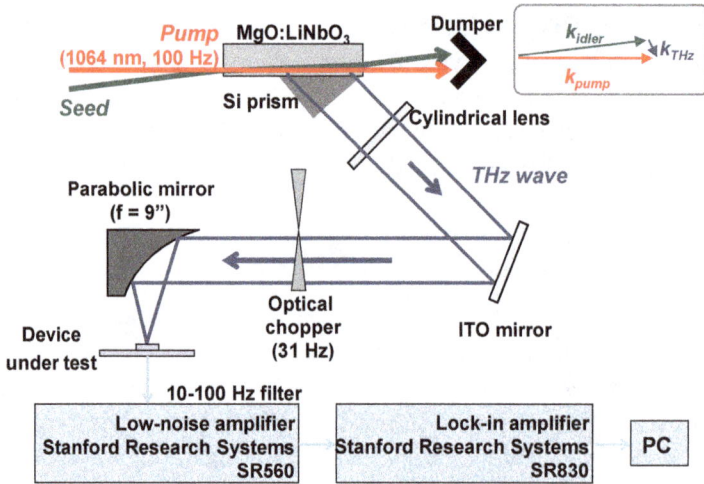

Fig. 6. Setup for responsivity measurements.

The measured responsivity of the fabricated imaging circuit is plotted in Fig. 7. The responsivity of the pixel was calculated from detected output voltage divided by available power to the antenna. The peak responsivity at 0.915 THz is 51.9 kV/W. The double-peak property of the measured responsivity is under investigation. The detector only draws about 3 µA from a power supply (V_{dd}) of 1.5 V. That means the detector has extremely low power consumption and is thus well suited for constructing a large-size array of imaging detectors.

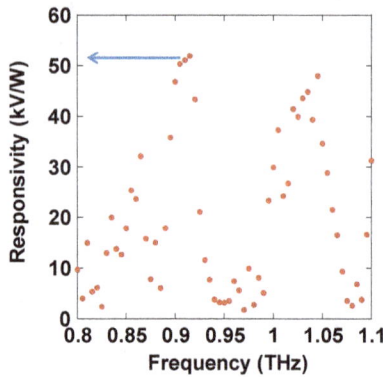

Fig. 7. Measured responsivity.

At a modulation frequency of 31 Hz, output noise level is 18.6 $\mu V/Hz^{1/2}$ (in Fig. 8), and measured NEP is 358 $pW/Hz^{1/2}$. NEP can be decreased by increasing sampling rate. Since the measured imaging array excludes an output buffer, the operation bandwidth of the detector with the extremely small current is degraded by the parasitic capacitances in the measurement equipment. If the parasitic capacitances of the measurement equipment are taken into account, the simulated gain well matches the measured gain (simulation A in Fig. 8). The operation bandwidth in the imaging circuit can be increased (corresponding to simulation B in Fig. 8). By increasing the sampling rate, NEP of 42 pW/Hz$^{1/2}$ is expected to be obtained at 100 kHz.

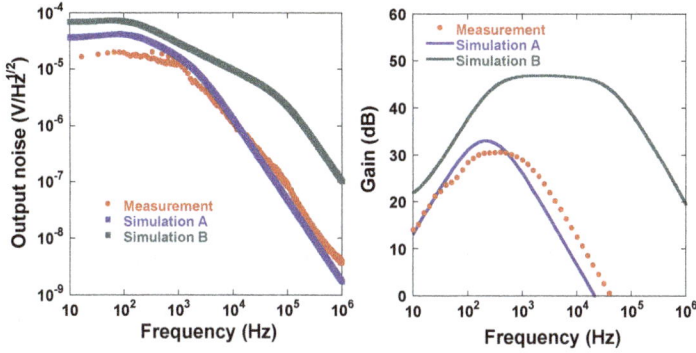

Fig. 8. Simulated and measured output noise and gain. Simulation A (left panel): with taking into account the parasitic capacitances in the measurement equipment (BNC cable and FRA). Simulation B (right panel): operation in the imaging circuit.

5. Conclusion

A terahertz imaging detector based on 180-nm CMOS process technology was designed and fabricated. The detector achieves a responsivity of 51.9 kV/W at 0.915 THz and a NEP of 358 pW/Hz$^{1/2}$ at a modulation frequency of 31 Hz. The performance of the detector is compared with recently developed ones in Table 1. According to the table, the proposed detector outperforms the existing detectors in terms of measured responsivity and power consumption. The detector's large NEP is mainly due to its low modulation frequency. And it can achieve a NEP of 42 pW/Hz$^{1/2}$ by increasing its modulation frequency to 100 kHz.

Table 1. Performance comparison.

Reference	Technology	Responsivity (kV/W)	NEP (pW/Hz$^{1/2}$)	Bias current (μA)	Chopping frequency
[14]	65 nm	0.8 @1 THz	66	-	1 kHz
[15]	130 nm	3.4 @0.82 THz	28	10	1 MHz
[16]	Schottky diode (130 nm)	0.273 @0.86 THz	42	20	1 MHz
This work	180 nm	51.9 @0.915 THz	358 42 (estimated)	3	31 Hz 100 kHz

Acknowledgments

This work was supported by the Ministry of Internal Affairs and Communications of Japan / Strategic Information and Communications R&D Promotion Programme (MIC/SCOPE) #151301001 and the VLSI Design and Education Center (VDEC), University of Tokyo in collaboration with Cadence Design System, Inc. and Keysight Technologies Japan, Ltd.

References

1. K. Kawase, Y. Ogawa, Y. Watanabe, and H. Inoue, "Non-destructive terahertz imaging of illicit drugs using spectral fingerprints", *Opt. Exp.* **11** (2003) 2549–2554.
2. M. C. Kemp, P. F. Taday, B. E. Cole, J. A. Cluff, A. J. Fitzgerald, and W. R. Tribe, "Security applications of terahertz technology", *Proc. SPIE* **5070**, *Terahertz for Military and Security Applications*, (2003) 44–52.
3. B. Zhu, Y. Chen, K. Deng, W. Hu, and Z. S. Yao, "Terahertz science and technology and applications", *PIERS Proc.*, Beijing, China, March 2009, pp. 1166–1170.
4. A. Hellicar, J. Du, N. Nikolic, L. Li, K. Greene, N. Beeton, S. Hanham, J. Kot, and G. Hislop, "Development of a terahertz imaging system", *2007 IEEE Antennas and Propagation Int. Symp.*, Honolulu, HI, USA, June 2007, pp. 5535-5538.
5. J. L. Hesler and T. W. Crowe, "Responsivity and noise measurements of zero-bias Schottky diode detectors", *Proc. 18th Int. Symp. on Space Terahertz Technology*, Pasadena, CA, USA, Mar. 2007, pp. 89–92.
6. P. Helisto, A. Luukanen, L. Gronberg, J. S. Penttila, H. Seppa, H. Sipola, C. R. Dietlein, and E. N. Grossman, "Antenna-coupled microbolometers for passive THz direct detection imaging arrays", *Proc. 2006 European Microwave Integrated Circuits Conf.*, Manchester, UK, Sep. 2006, pp. 35–38.
7. L. Samoska, "An overview of solid-state integrated circuit amplifiers in the submillimeter-wave and THz ragime", *IEEE Trans. Terahertz Sci. Technol.* **1** (2011) 9–24.
8. S. Eminoglu, M. Tanrikulu, and T. Akin, "A low-cost 128 x 128 uncooled infrared detector array in CMOS process", *J. Microelectromech. Syst.* **17** (2008) 20–30.
9. E. Ojefors, A. Lisauskas, H. G. Roskos, and U. R. Pfeiffer, "Terahertz imaging detectors in CMOS technology", *J. Infrared Milim. Terahertz Waves* **30** (2009) 1269–1280.
10. R. A. Hadi, H. Sherry, J. Grzyb, Y. Zhao, W. Forster, H. M. Keller, A. Cathelin, A. Kaiser, and U. R. Pfeiffer, "A 1 k-pixel video camera for 0.7-1.1 terahertz imaging applications in 65-nm CMOS", *IEEE J. Solid-State Circ.* **47(12)** (2012) 2999–3012.
11. P. E. Allen and D. R. Holberg, *CMOS Analog Circuit Design Second Ed.*, Oxford University Press, USA, 2002, pp. 72–79.
12. T. Wada, M. Ikebe, and E. Sano, "60-GHz, 9 µW wake-up receiver for short-range wireless communications", *Proc. Eur. Solid-State Circuits Conf. (ESSCIRC)*, Bucharest, Romania, Sep. 2013, pp. 383–386.
13. S. Hayashi, K. Nawata, T. Taira, J. Shikata, K. Kawase, and H. Minamide, "Ultrabright continuously tunable terahertz-wave generation at room temperature", *Sci. Rep.* **4** (2014) 5045.
14. R. Hadi, H. Sherry, J. Grzyb, N. Baktash, Y. Zhao, E. Öjefors, A. Kaiser, A. Cathelin, and U. Pfeiffer, "A broadband 0.6 to 1 THz CMOS imaging detector with an integrated lens", *Proc. 2011 IEEE Microwaves Symp.*, Baltimore, MD, USA, June 2011, pp. 1–4.

15. D. Y. Kim, S. Park, R. Han, and Kenneth K. O, "820-GHz imaging array using diode-connecter NMOS transistors in 130-nm CMOS," *Dig. Tech. Papers 2013 IEEE VLSI Circuits Symp.*, Kyoto, Japan, June 2013, pp. 12–13.

16. R. Han, Y. Zhang, Y. Kim, D. Y. Kim, H. Shichijo, E. Afshari, and Kenneth K. O, "Active terahertz imaging using Schottky diodes in CMOS: array and 860-GHz pixel", *IEEE J. Solid-State Circ.* **48(10)** (2013) 2296–2308.

Plasma Instability of 2D Electrons in a Field Effect Transistor with a Partly Gated Channel

Aleksandr S. Petrov[1], D. Svintsov[2], and M. Rudenko[3]

Laboratory of 2D Materials' Optoelectronics, MIPT, Dolgoprudny, Russia
[1]aleksandr.petrov@phystech.edu
[2]svintcov.da@mipt.ru
[3]mike.rudenko@gmail.com

V. Ryzhii

Research Institute of Electrical Communication, Tohoku University, Sendai, Japan
v-ryzhii@riec.tohoku.ac.jp

M. S. Shur

Department of Electrical, Computer, and System Engineering and Department of Physics,
Applied Physics, and Astronomy, Rensselaer Polytechnic Institute, Troy, New York 12180, USA
shurm@rpi.edu

We predict the instability of plasma waves excited by a DC current in the field-effect transistors (FETs) with a partly gated channel. The excitation of plasma waves is due to amplified reflection from the boundary of the gated and ungated regions. The boundary also supports the turbulent edge modes whose increment strongly depends on the ratio of the carrier densities in the two FET regions.

Keywords: 2d electron gas; plasma instabilities.

1. Introduction

The interest for plasma instabilities in field-effect transistor (FET) channels under dc current is growing continuously due to their possible applications for the generation of THz radiation.[1,2] In this regard, the Dyakonov-Shur plasma instability[3] is of particular interest as the spectrum of the excited oscillations is resonant. Its mechanism is based on (1) the difference of plasma wave velocities propagating upstream and downstream to the dc current and (2) the ideal ac current reflection at the drain side of the transistor. Upon the reflection from the drain, the magnitude of small perturbations of electron density increases by $(s+u)/(s-u)$, where u is the electron flow velocity and s is the plasma wave velocity.

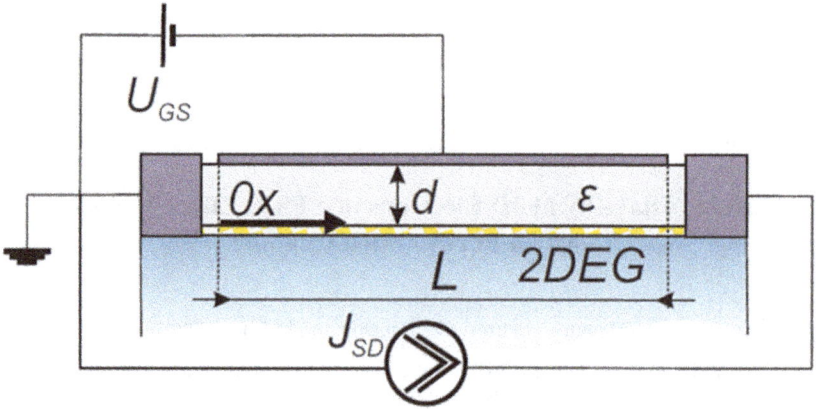

Fig. 1. Schematic view of a FET with 2d channel where the Dyakonov-Shur plasma instability is supposed to develop.

To provide the required boundary condition for the Dyakonov-Shur instability, one has to support a constant dc current between the source and drain contacts (Fig. 1). Later, it was shown[8] that the precise form of the boundary conditions used above is not crucial for the instability and that a positive growth increment can be obtained for arbitrary terminating source and drain impedances provided that the imaginary part of the source impedance is greater than the imaginary part of the drain impedance. However, all FET contacts are commonly connected to the voltage sources. In this regard, it was argued[4,8] that the condition of ideal ac current reflection can be approximately fulfilled for the transistors operating in the saturation mode. However, the possibility of instability growth under current saturation in fully gated channel has been proved neither analytically nor numerically.

With a direct numerical solution of hydrodynamic equations and Poisson equation in a gated FET connected to a voltage source, it is possible to show that the operation of a fully gated FET in the saturation mode does not provide the condition of the plasma instability. The results of such simulation are shown in Fig. 2(b). After a rapid switch-on of the drain voltage, the current oscillations at the drain side quickly decay, and the current reaches its steady-state value. On the contrary, if the drain is connected to a current source, the potential at the drain evolves toward a periodic oscillation [Fig. 2(a)].

2. Plasma Instability in FET with a Partly Gated Channel

In this work we show that plasma instability similar to that predicted by Dyakonov and Shur is feasible under realistic boundary conditions (fixed source and drain potentials) in an asymmetric transistor with a partly gated channel (Fig. 3). The mechanism of instability does not rely on the transit-time effects occurring under current saturation in the FET, in contrast to the instability predicted in Ref. 5. It is also different from the absolute negative resistance effect predicted in Ref. 9 for a ballistic FET with the threshold voltage varying as a function of the coordinate in the channel.

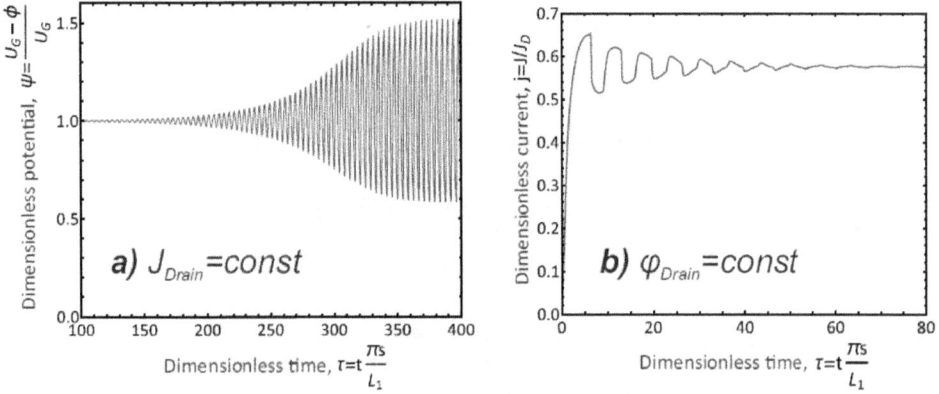

Fig. 2. Results of the numerical modeling of plasma instabilities in a FET channel under various boundary conditions at the drain: a) constant current and b) constant voltage bias. In the case (a) at $t = 0$ a constant current source is switched to the drain. In the case (b) at $t = 0$ a voltage exceeding the saturation voltage is applied at the drain. The growth of plasma oscillations predicted by Dyakonov-Shur theory can be seen in (a). In the case (b) the transient plasma oscillations due to rapid switching of transistor decay. Parameters of the modelling: a) dimensionless drain current $j_D = J_D / (n_0 s) = 0.04$, $n_0 = C U_g / e$, dimensionless scattering rate $\gamma = L / (\tau s) = 0.01$, reciprocal Reynolds number $1 / \mathrm{Re} = 0.01$, where C is the capacitance per unit area, L – length of the gate, $1 / \tau$ – effective scattering rate, U_G - constant gate voltage; b) $\gamma = 0.1$, $1 / \mathrm{Re} = 10^{-4}$, drain potential $\varphi_D = U_G / 2$.

As the ratio of plasmon wavelengths in ungated and gated regions is large, the plasma wave reflection coefficient from their boundary is close to unity in the absence of a DC current. The effect is similar to the Fresnel reflection from the medium with a higher refractive index. Under the passage of a DC current, the reflection of plasma oscillations from the gated/ungated boundary entails their growth.

We solve the linearized set of hydrodynamic equations and Poisson's equation for the structure shown in Fig. 3 under boundary conditions of: (1) a fixed potential on source and drain contacts and (2) the continuity of the potential and ac current across the gated/ungated boundary. This procedure leads to the following dispersion equation for plasma waves:

$$\frac{s_1 + u_1}{s_1 - u_1} \frac{1 - 2ik_+ d\,\mathrm{ctg}\left[\left(\kappa_+ - \kappa_-\right)\dfrac{L_2}{2}\right]}{1 - 2ik_- d\,\mathrm{ctg}\left[\left(\kappa_+ - \kappa_-\right)\dfrac{L_2}{2}\right]} = -e^{-i(k_+ - k_-)L_1}, \tag{1}$$

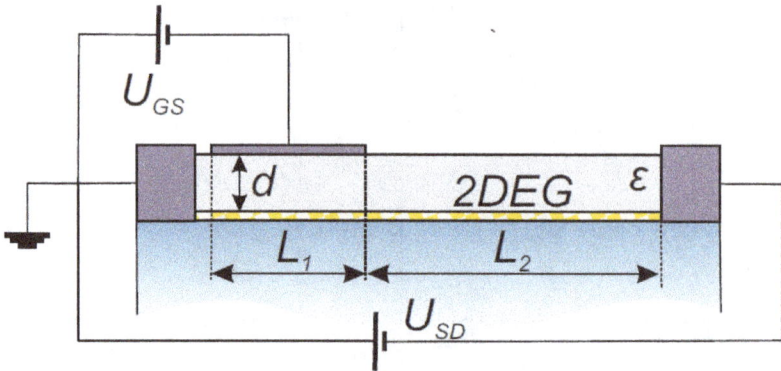

Fig. 3. Scheme of a FET with a partly gated channel.

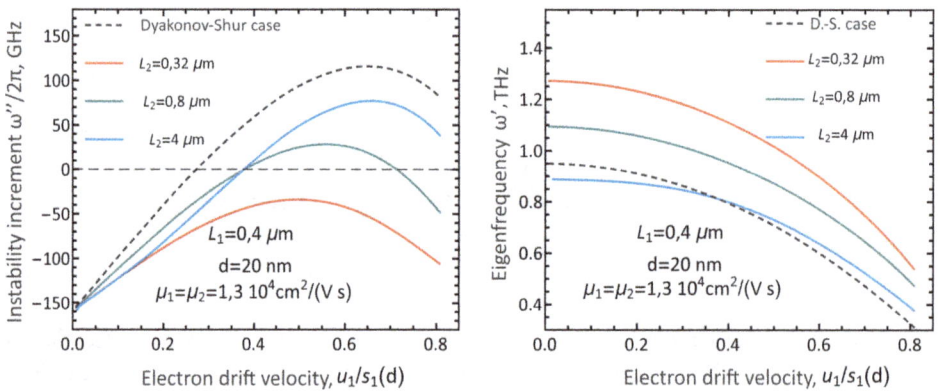

Fig. 4. Dependencies of the plasma wave increment and eigenfrequencies on carrier drift velocity at different lengths of the ungated part of the transistor channel.

where $k_{\pm}(\omega)$ and $\kappa_{\pm}(\omega)$ are the wave vectors of the 'upstream' (minus sign) and 'downstream' (plus sign) waves in the gated and ungated parts of the channel, respectively, the left-hand side of the equation being just the reflection coefficient from the gated/ungated boundary.

The numerical solution of Eq. (1) is presented in Fig. 4. It can be seen that even at relatively low mobilities of the order 10_4 cm^2/(V·s) (which correspond to GaAs at room temperature) the self-excitation of plasma oscillations is possible and the increment is comparable with that of Dyakonov-Shur instability with ideal current source (dashed line in Fig. 4). The instability increment is more pronounced at larger lengths of the ungated region; at $L_2 \to 0$ one recovers the case of constant voltage source is applied to the drain of fully gated FET, and no instability occurs. The real part of the eigenfrequency only slightly depends on L_2.

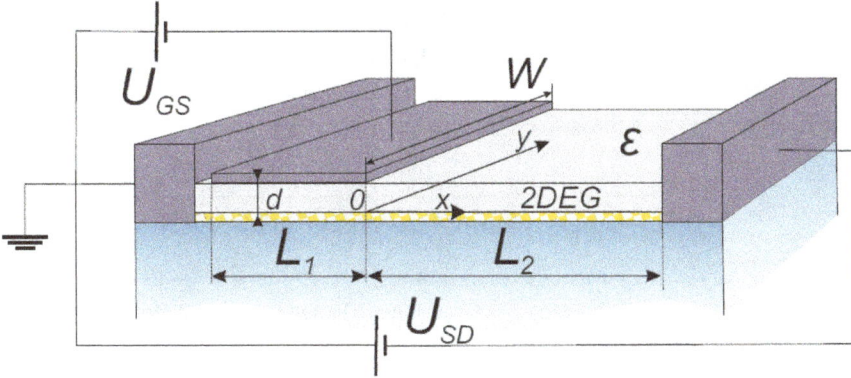

Fig. 5. Scheme of a FET with a partly gated channel in 'extended' geometry for the study of plasmons propagating along the gated/ungated boundary.

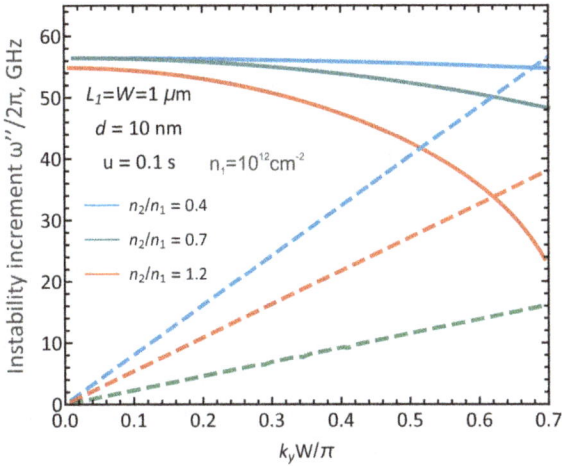

Fig. 6. Dependence of the plasma wave increment of the bulk (solid lines) and edge (dashed lines) modes on the y-component of the plasma wave vector at different ratio of carrier concentrations in two sections of a FET.

3. Edge Modes

Another type of unstable modes is revealed in the Dyakonov-Shur FET if one considers the waves propagating along the transistor contacts.[6] These modes are localized near the drain side of the channel and are evanescent in the Ox direction (edge modes). It was shown that the increment of such modes can be much higher than that of the 'bulk' modes, for this reason the nonlinear response of FET might be largely governed by the excitation of the edge waves.

Taking this into account, we expect to find similar unstable modes localized at the gated/ungated boundary by extending our geometry to the two dimensions (Fig. 5). The dispersion equation is obtained with a procedure similar to that of Sec. II, and the new type of solutions corresponding to the edge modes appear once the carrier densities in gated and ungated regions are not equal.

For the bulk modes, the increment decreases at large k_y (Fig. 6, solid lines), in contrast to the edge modes (Fig. 6, dashed lines). The increment of the edge modes depends on the ratio of carrier densities in the two sections of the FET. This effect can be studied analytically using the dispersion relation of the edge modes:

$$\frac{k_y}{\omega'' - k_y u_1} e^{-2k_y L_1} = \frac{k_y}{\omega'' + k_y u_1} + \frac{n_2}{n_1} \frac{k_y}{\omega'' - k_y u_1 n_1 / n_2}, \tag{2}$$

where n_1 and n_2 are the carrier densities in the gated and ungated regions, respectively. When these are close to each other, the increment becomes proportional to a vanishing exponent with respect to L_1, $\omega'' \approx k_y u_1 e^{-2k_y L_1} / 2 - e^{(-2k_y L_1)}$. However, when the densities differ considerably, the increment grows with their relative difference: $\omega'' = k_y u_1 (n_1 - n_2) / n_2$. This leads us to the conclusion that for the intense growth of the edge modes not only the structure asymmetry, but also the carrier density asymmetry is required. This conclusion is consistent with the numerical simulations[10] that revealed that the edge modes are not excited in the device with a symmetrical carrier distribution. On the other hand, by proper tuning of carrier density in the two regions, one can 'switch off' the edge modes. This can be necessary for some applications, as the development of edge instabilities leads to a broad spectrum of the emitted radiation.[6]

4. Conclusion

We have shown that the plasma wave self-excitation upon reflection from the gated/ungated boundary in the field-effect transistors is possible. The mechanism of instability does not rely on the transit-time effects in the ungated region, and thus the plasma wave self-excitation can occur even in the pre-saturation mode of the FET for sufficiently large carrier mobilities. Such instability can occur in the FETs where the current saturation is hardly achievable, particularly, in graphene transistors.[7]

We have also shown that the new type of unstable modes is FETs with a partly gated channel is localized at the gated/ungated boundary. Their increment can exceed that of bulk modes, especially at a large ratio of carrier densities in the gated and ungated regions.

Acknowledgements

This work was supported by the grant No. 16-19-10557 of the Russian Science Foundation.

References

1. A. El Fatimy, N. Dyakonova, Y. Meziani, T. Otsuji, W. Knap, S. Vandenbrouk, K. Madjour, D. Théron, C. Gaquiere, M. A. Poisson, S. Delage, P. Prystawko, and C. Skierbiszewski, "AlGaN/GaN high electron mobility transistors as a voltage-tunable room temperature terahertz sources", *J. Appl. Phys.* **107** (2010) 024504.
2. T. Otsuji, T. Watanabe, S. A. Boubanga Tombet, A. Satou, W. M. Knap, V. V. Popov, M. Ryzhii, and V. Ryzhii, "Emission and detection of terahertz radiation using two-dimensional

Electrons in III–V semiconductors and graphene", *IEEE Trans. Terahertz Sci. Technol.* **3** (2013) 63–71.

3. M. Dyakonov and M. S. Shur, "Shallow water analogy for a ballistic field effect transistor: New mechanism of plasma wave generation by dc current", *Phys. Rev. Lett.* **71** (1993) 2465.

4. W. Knap, J. Lusakowski, T. Parenty, S. Bollaert, A. Cappy, V. V. Popov, and M. S. Shur, "Terahertz emission by plasma waves in 60 nm gate high electron mobility transistors", *Appl. Phys. Lett.* **84** (2004) 2331.

5. V. Ryzhii, A. Satou, and M. S. Shur, "Transit-time mechanism of plasma instability in high electron mobility transistors", *Phys. Status Solidi* **202** (2005) R113–R115.

6. M. Dyakonov, "Boundary instability of a two-dimensional electron fluid", *Semiconductors* **42** (2008) 984–988.

7. I. Meric, M. Y. Han, A. F. Young, B. Ozyilmaz, P. Kim, and K. L. Shepard, "Current saturation in zero-bandgap, top-gated graphene field-effect transistors", *Nat. Nanotech.* **3** (2008) 654–659.

8. M. Dyakonov and M. S. Shur, "Plasma wave electronics for terahertz applications", in *Terahertz Sources and Systems* **27**, eds. R. E. Miles, P. Harrison, and D. Lippens, Dordrecht/Boston/London, 2001, pp. 187–207.

9. M. Dyakonov and M. Shur, "Current instability and plasma wave generation in ungated two dimensional electron layers", in *Frontiers in Electronics, Selected Topics in Electronics and Systems* **41**, eds. H. Iwai, Y. Nishi, M. S. Shur and H. Wong, World Scientific, Singapore 2006, pp. 443–451.

10. S. Rudin and G. Rupper, "Plasma instability and wave propagation in gate-controlled GaN conduction channels", *Jpn. J. Appl. Phys.* **52** (2013) 08JN25.

THz Spectroscopic Imaging of Chemicals Using IS-TPG

Kosuke Murate[1,2,*], Mikiya Kato[1], and Kodo Kawase[1,3]

[1]*Department of Electronic Engineering, Nagoya University, Furo-cho, Chikusa,
Nagoya, 464-8603, Japan*
*murate.kousuke@h.mbox.nagoya-u.ac.jp

[2]*Japan Society for the Promotion of Science, Kojimachi, Chiyoda,
Tokyo, 102-0083, Japan*

[3]*RIKEN, Aramaki-Aoba, Aoba, Sendai, 980-0845, Japan*

In 2003, we demonstrated a non-destructive terahertz spectroscopic imaging of illicit drugs hidden in envelopes using a widely tunable THz-wave parametric source, though its dynamic range at that time was less than four orders. Recently, we have realized more than nine orders of dynamic range with an evolved injection seeded THz-wave spectrometer. Now we can detect drugs under much thicker obstacles than before. In this report, we introduce the result of THz spectroscopic imaging of chemicals under thick envelopes and related topics; Enhanced tuning range of is-TPG up to 5 THz and >90 dB dynamic range THz spectroscopic system.

Keywords: is-TPG; spectroscopic imaging; nonlinear optics.

1. Introduction

Terahertz (THz) wavelengths lie between the microwave and infrared regions of the electromagnetic spectrum, and share the characteristics of both. As with radio waves, THz-wave can be partially transmitted through a wide variety of materials, including plastics, ceramics, and papers. As with infrared wave, THz-wave can be guided using mirrors and lenses. Besides this, a large number of materials have unique absorption characteristics (also termed fingerprints) at THz frequencies. For these reasons, THz-wave has a number of potential applications in biomedical, communication, non-destructive sensing, and imaging fields.[1]

For years, we have worked on the development of a high-power THz-wave source, based on parametric processes in a MgO:LiNbO$_3$ crystal in order to develop the high dynamic range THz-wave spectroscopic system for non-destructive testing.[2-4] In 2003, we reported the non-destructive detection of illicit drugs hidden in envelopes.[5] However, the dynamic range of THz wave parametric oscillator (TPO)[2,3] used at that time was below 4 orders, so we demonstrated screening and detection through a thin international mail envelope that was about 0.1 mm thick. Recently, the peak output power of injection seeded terahertz parametric generator (is-TPG) approached >50 kW[6] by introducing a new pump laser; a microchip Nd:YAG laser with shorter pulse width.[7,8] Moreover, highly sensitive THz-wave detection was possible using the principle of is-TPG.[6,9]

Therefore, we could develop the THz-wave spectroscopic system with more than nine orders of dynamic range by combination of such source and detector.[10] Using this high dynamic range spectroscopic system, now we can detect chemicals under much thicker obstacles than before.

In this report, firstly, we introduce the improvement of is-TPG spectroscopic system; i) Enhanced tuning range up to 5 THz, ii) >90 dB dynamic range THz detection using near infrared detector. Secondly, we explain about THz spectroscopic imaging of chemicals using is-TPG; iii) Comparison between is-TPG spectrometer and THz-TDS, and iv) THz Spectroscopic imaging.

2. THz-Wave Emission and Detection by is-TPG

Figure 1 shows the experimental setup for the emission and detection of THz wave using is-TPG. Microchip Nd:YAG laser[7,8] and continuous wave tunable external cavity diode laser were used as pump and seed beam respectively. The pump beam was divided by beam splitter and part of it along with seed beam irradiated the MgO:LiNbO₃ crystal to emit the tunable THz-wave.[10] In this process, non-linear phase matching condition must be satisfied for the efficient generation of broadband THz wave. Therefore, an achromatic optics consisting of grating and pair of lenses was implemented for the dispersion compensation in emitter crystal. The emitted THz wave was then focused to another MgO:LiNbO₃, which is also pumped by pump beam. Finally, the THz waves were detected based on nonlinear optical wavelength conversion,[6,9] where we measured the idler beam intensity using infrared detector. Since the idler beam was angle dispersive, we implemented pyroelectric detector mounted in computer controlled linear stage.

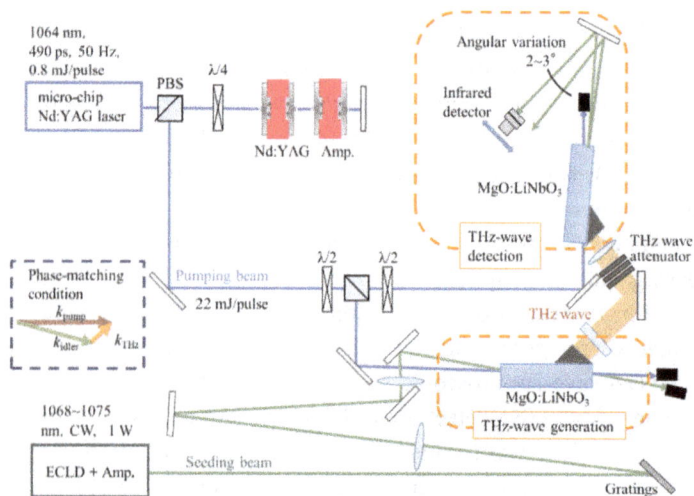

Fig. 1. Experimental setup of is-TPG spectroscopic system.

3. Improvement of is-TPG Spectroscopic System

3.1. *Enhanced tuning range up to 5 THz*

The output peak power of is-TPG was improved drastically as explained in previous paragraph, however, the tuning range of is-TPG was still limited less than 3 THz due to the strong absorption loss inside the MgO:LiNbO$_3$ crystal: the absorption increased with the frequency.[11–13] Inside a MgO:LiNbO$_3$ crystal, a THz-wave is generated across the pump beam via excitation from the pump pulse under non-collinear phase-matching conditions. Figure 2(a) shows that, at the center core region of the pump beam, the THz-wave is well amplified by the pump beam with the gain being larger than the absorption loss. In contrast, as the THz-wave propagates outside of the core region of the pump beam, the THz-wave is no longer amplified, because the gain is less than the absorption loss. Thus, it is more difficult for a higher frequency THz-wave to reach to the THz-wave exit surface (y-surface).

Here, we slightly inclined the MgO:LiNbO$_3$ crystal so that a portion of the pump beam was internally reflected at the y-surface, as shown in Fig. 2(b). In this configuration, it became easier for a higher frequency THz-wave to reach to the y-surface, because the center core region of the pump beam was closer to that surface. As a result, the tuning range was expanded to a range of 0.6–5 THz as shown in Fig. 3, resulting in band tunability of almost one order.[14]

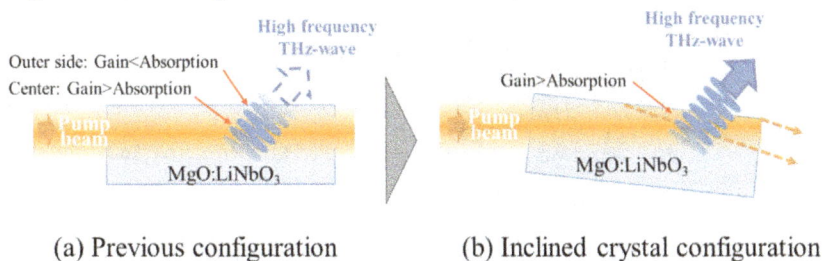

High frequency THz-wave

Outer side: Gain<Absorption
Center: Gain>Absorption

Pump beam

MgO:LiNbO$_3$

High frequency THz-wave

Gain>Absorption

Pump beam

MgO:LiNbO$_3$

(a) Previous configuration   (b) Inclined crystal configuration

Fig. 2. Inclination of the crystal to reduce terahertz (THz)-wave absorption loss in the crystal: (a) previous configuration and (b) inclined crystal configuration (the seed beam is not shown here for simplicity of the explanation). Experimental setup of is-TPG spectroscopic system.

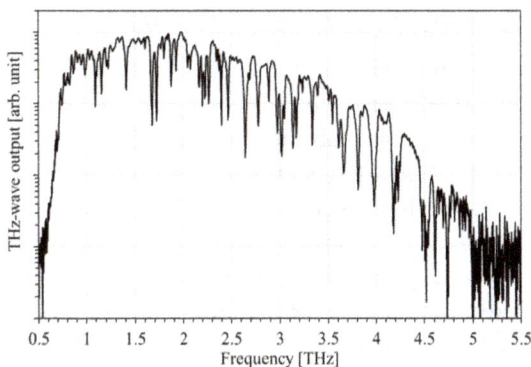

Fig. 3. Achievement of tuning range expansion of the is-TPG by optimization of the crystal angle orientation.

3.2. *>90 dB dynamic range THz detection using near infrared detector*

In the detection section of our system, THz-wave was converted back into near infrared beam by nonlinear optical wavelength conversion.[6,9] Since the detection methodologies in optical frequency region are well established and high sensitive near infrared detectors are readily available, we were able to measure the extremely small THz-wave output by measuring the power of converted near infrared beam. In order to measure the dynamic range of our system, the energy of the emitted THz-wave was varied for 10 orders of magnitude using THz wave attenuators. At the frequency of 1.5 THz, we were able to achieve the dynamic range of > 90 dB using commercially available near infrared photo detector as shown in Fig. 4. This system has potential applications in non-destructive sensing and imaging of a wide variety of materials.[15,16]

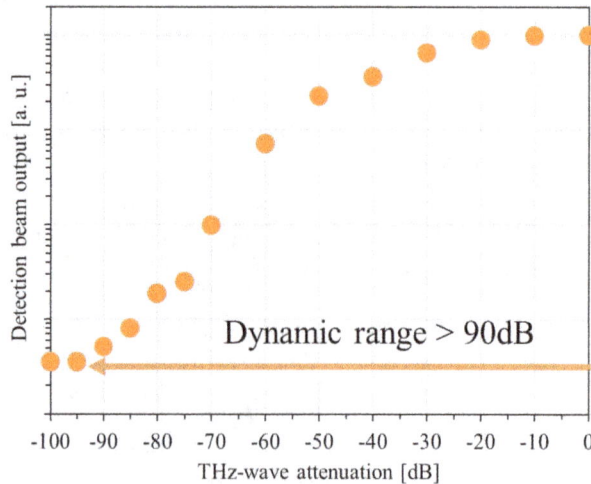

Fig. 4. The dynamic range of the system at 1.5 THz.

4. THz Spectroscopic Measurement

4.1. *Comparison between is-TPG spectrometer and THz-TDS*

An is-TPG is monochromatic with a widely tunable THz wave source and can obtain THz spectra directly over a relatively wide detection area. Therefore, the spectra from the contents contained in covering materials that refract, diffract, or scatter THz waves can be measured using is-TPG. Recent is-TPG research has resulted in a significant increase in power output and the highly sensitive detection of THz-waves[6] as explained in previous paragraph. Studies have also developed a high dynamic range spectrometer using an is-TPG.[10] In this study, in order to evaluate the performance of is-TPG in mail inspection,[5] we measured the transmission spectra of various chemicals under covering materials using an is-TPG spectrometer, and compared the spectra with those measured by terahertz time-domain spectroscopy (THz-TDS).

The samples used in this study were shaped to appear as illicit drugs concealed in international mail. To this end, we enclosed three powdered saccharides — maltose, glucose, and fructose — in plastic bags that were approximately 0.5 mm thick, including the powder. Then, the bags were placed in EMS envelopes which are used for international mail, and bubble wraps. Each saccharide has several absorption peaks from 1.0 to 2.5 THz, with maltose showing characteristic absorption at 1.2, 1.6, and 2.0 THz, glucose at 1.4 and 2.1 THz, and fructose at 1.7 and 2.1 THz.[17]

Figure 5 shows the transmission spectrum of each sample measured using is-TPG and THz-TDS. Using is-TPG, all of the absorption peaks except for the maltose absorption peak at 1.2 THz could be identified clearly. By contrast, using THz-TDS, the absorption peaks below 1.6 THz for each sample could be identified, while the absorption peaks at higher frequencies could not be distinguished from background noise. Similar results were seen using complex shape covering materials, such as corrugated cardboard. These results demonstrated that is-TPG spectroscopic system was effective at identifying chemical compounds within covering substances, i.e., in detecting illicit drugs concealed in mail.

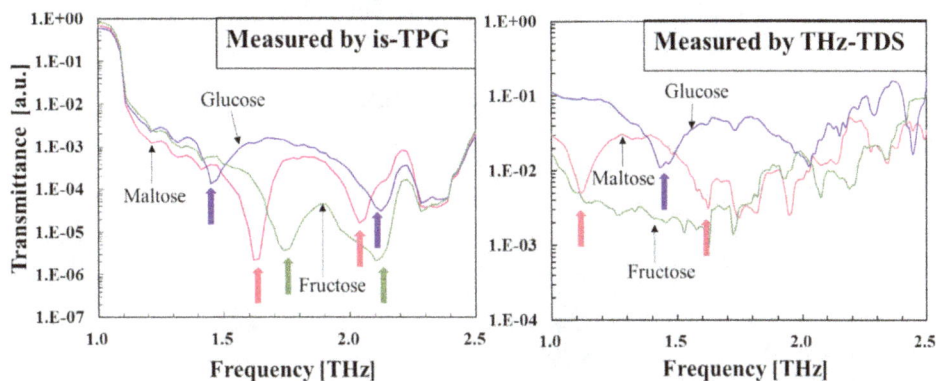

Fig. 5. Transmission spectra of powdered saccharides in EMS envelops and bubble wraps.

4.2. *THz spectroscopic imaging*

Now we can detect drugs under much thicker obstacles than before using evolved is-TPG spectroscopic imaging system[10] as explained in previous paragraph. The samples used in this study were three saccharides: maltose, glucose, and fructose as same as spectroscopic measurement. During the sample preparation, particle sizes for each powder were selected to be 30-130 μm and the powders were placed in plastic bags to prepare 1-mm-thick samples. The size of each sample was approximately 10 mm × 10 mm. The three saccharide samples were positioned in the order of maltose, glucose, and fructose, as shown in right side of Fig. 6. To create a denser wrapping, three types of covering material were used: EMS envelopes, corrugated board, and bubble wraps. Two EMS envelopes, two pieces of corrugated board, and four sheets of bubble wraps were placed on the front and back of the saccharide samples as shown in Fig. 6. The thickness after affixing the coverings was approximately 23 mm, much thicker than the 0.1-mm-thick envelopes used in the 2003 study.[5]

The result of spectroscopic imaging through a thick shielding material is shown in Fig. 7(a), which illustrates that the three saccharide samples were identified separately despite the thick wrapping materials. Moreover, the density of the white color captures the spatial distribution of the powders. Figure 7(b) shows the one-pixel absorption spectrum of each saccharide sample showing the absorption peaks, thereby correctly identifying each saccharide sample.[15]

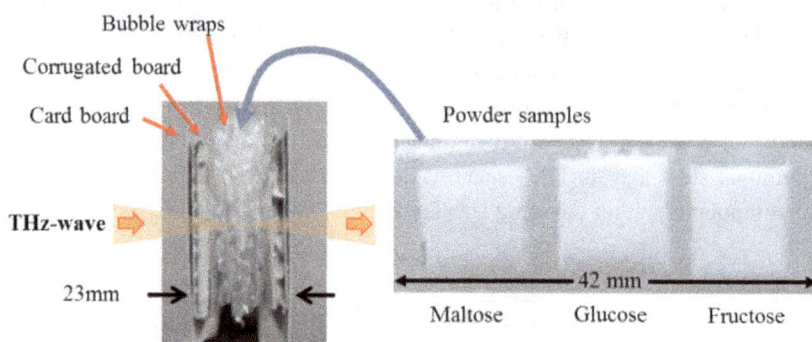

Fig. 6. Sample preparation. Three saccharide samples were shielded by 23 mm thick covering materials.

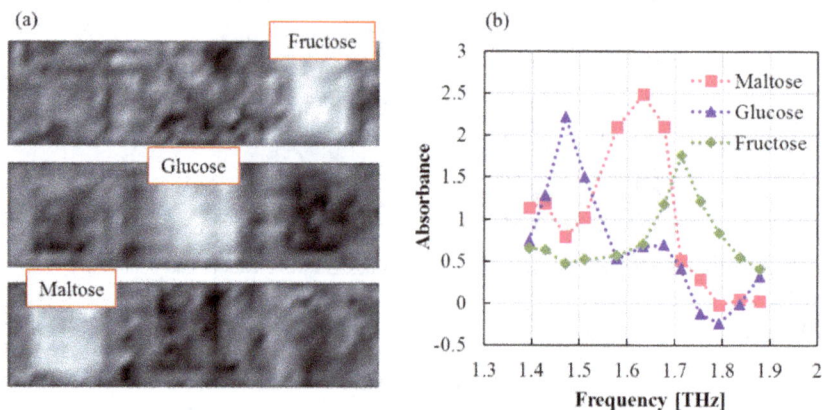

Fig. 7. (a) THz spectroscopic imaging of powder samples in two cardboards + four bubble wraps + two corrugated boards. (b) One-pixel absorption spectrum of each saccharide sample.

5. Conclusion

In this report, we explained about improvement of is-TPG spectrometer and THz spectroscopic imaging of chemicals using is-TPG. The tuning range was expanded to a range of 0.6 – 5 THz, and the dynamic range reached to more than nine orders. A THz spectral imaging system was constructed that used such THz measurement system, and successfully identified images of saccharides through wrapping material much thicker than those used in the 2003 trials. Using this system, spectral imaging is possible for not only saccharides but also cases where illegal drugs and multiple reagents are mixed together. This method can be used to inspect mail, check prescription accuracy at pharmacies, inspect pharmaceutical manufacturing processes, and determine illegal possession of stimulants and explosives.

Acknowledgments

The authors appreciate the fruitful discussions with Prof. S. R. Tripathi of Nagoya University, Dr. H. Minamide, Dr. S. Hayashi, Dr. K. Nawata of Riken and Prof. T. Taira of IMS. This work was partially supported by JSPS KAKENHI Grant Number 25220606, 15J04444.

References

1. M. Tonouchi, "Cutting-edge terahertz technology", *Nat. Photon.* **1** (2007) 97–105.
2. K. Kawase, M. Sato, T. Taniuchi, and H. Ito, "Coherent tunable THz-wave generation from LiNbO₃ with monolithic grating coupler", *Appl. Phys. Lett.* **68** (1996) 2483.
3. K. Kawase, J. Shikata, and H. Ito, "Terahertz wave parametric source", *J. Phys. Appl. Phys.* **35** (2002) R1.
4. K. Kawase, H. Minamide, K. Imai, J. Shikata, and H. Ito, "Injection-seeded terahertz-wave parametric generator with wide tunability", *Appl. Phys. Lett.* **80** (2002) 195.
5. K. Kawase, Y. Ogawa, Y. Watanabe, and H. Inoue, "Non-destructive terahertz imaging of illicit drugs using spectral fingerprints", *Opt. Exp.* **11** (2003) 2549–2554.
6. S. Hayashi, K. Nawata, T. Taira, J. Shikata, K. Kawase, and H. Minamide, "Ultrabright continuously tunable terahertz-wave generation at room temperature", *Sci. Rep.* **4** (2014) 5045.
7. H. Sakai, H. Kan, and T. Taira, "> 1 MW peak power single-mode high-brightness passively Q-switched Nd³⁺: YAG microchip laser", *Opt. Exp.* **16** (2008) 19891–19899.
8. T. Taira, "Domain-controlled laser ceramics toward giant micro-photonics [Invited]", *Opt. Mater. Exp.* **1** (2011) 1040–1050.
9. R. Guo, S. Ohno, H. Minamide, T. Ikari, and H. Ito, "Highly sensitive coherent detection of terahertz waves at room temperature using a parametric process", *Appl. Phys. Lett.* **93** (2008) 21106.
10. K. Murate *et al.*, "A high dynamic range and spectrally flat terahertz spectrometer based on optical parametric processes in LiNbO₃", *IEEE Trans. Terahertz Sci. Technol.* **4** (2014) 523–526.
11. D. R. Bosomworth, "The far infrared optical properties of LiNbO₃", *Appl. Phys. Lett.* **9** (1966) 330.
12. A. de Bernabé, C. Prieto, and A. de Andrés, "Effect of stoichiometry on the dynamic mechanical properties of LiNbO₃", *J. Appl. Phys.* **79** (1996) 143.
13. L. Pálfalvi, J. Hebling, J. Kuhl, Á. Péter, and K. Polgár, "Temperature dependence of the absorption and refraction of Mg-doped congruent and stoichiometric LiNbO₃ in the THz range", *J. Appl. Phys.* **97** (2005) 123505.
14. K. Murate, S. Hayashi, and K. Kawase, "Expansion of the tuning range of injection-seeded terahertz-wave parametric generator up to 5 THz", *Appl. Phys. Exp.* **9** (2016) 82401.
15. M. Kato, S. R. Tripathi, K. Murate, K. Imayama, and K. Kawase, "Non-destructive drug inspection in covering materials using a terahertz spectral imaging system with injection-seeded terahertz parametric generation and detection", *Opt. Exp.* **24** (2016) 6425.

16. S. R. Tripathi, Y. Sugiyama, K. Murate, K. Imayama, and K. Kawase, "Terahertz wave three-dimensional computed tomography based on injection-seeded terahertz wave parametric emitter and detector", *Opt. Exp.* **24** (2016) 6433.
17. RIKEN and NICT, "THz-database." [Online]. Available: http://thzdb.org/.

Imaging and Gas Spectroscopy for Health Protection in Sub-THz Frequency Range

W. Knap[*], D. B. But, D. Couquillat, and N. Dyakonova

Laboratoire Charles Coulomb, University of Montpellier & CNRS, France,
Montpellier 34980, France
[*]knap@univ-montp2.fr

M. Sypek[†] and J. Suszek

Faculty of Physics, Warsaw University of Technology,
Warsaw, 00-662 Poland
[†]sypek@if.pw.edu.pl

E. Domracheva, M. Chernyaeva, and V. Vaks[‡]

Terahertz Spectrometry Department, Institute for Physics of Microstructures,
Nizhny Novgorod, 603950, Russia
[‡]vax@ipmras.ru

K. Maremyanin and V. Gavrilenko[§]

Department for physics of semiconductors, Institute for Physics of Microstructures,
Nizhny Novgorod, 603950, Russia
[§]gavr@ipmras.ru

C. Archier[**] and B. Moulin

T-Waves Technologies Montpellier,
Montpellier, 34095, France
[**]christophe.archier@t-waves-technologies.com

G. Cywinski[††], I. Yahniuk, and K. Szkudlarek

Institute of High Pressure Physics of the Polish Academy of Sciences
Warsaw, 01-142, Poland
[††]gc@unipress.waw.pl

An overview of main results concerning THz detection related to plasma nonlinearities in nanometer field effect transistors is presented. In particular the physical limits of the responsivity, speed and the dynamic range of these detectors are discussed. As a conclusion, we will present applications of the FET THz detectors for construction of focal plane arrays. These arrays, together with in purpose developed diffractive 3D printed optics lead to construction of the demonstrators of the fast postal security imagers and nondestructive industrial quality control systems. We will show also first results of FET based imaging that uses for contrast not only usual THz radiation amplitude, but also the degree of its circular polarization. Sub-THz high resolution gas spectroscopy is shown to be a powerful means to diagnose various diseases via exhaled breath analysis.

Keywords: Terahertz FET detectors; focal plane arrays; exalted breath analysis.

1. Introduction

We present an overview of some recent results concerning THz detection related to plasma nonlinearities in nanometer field effect transistors.[1,2] The subjects were selected in a way to show physics related limitations and advantages of nanometer field effect transistors (FETs) working as terahertz detectors. We address the problems like limits of the responsivity and its temperature dependence,[3] helicity sensitive detection[4] and nonlinear/saturation response at high incident power.[5] The results will be discussed in view of the physical limitations of FET based THz detectors in view of their application for terahertz imaging.[6–8] We show that FET based arrays, together with in purpose developed diffractive 3D printed optics lead to construction of the demonstrators of the fast postal security imagers and nondestructive industrial quality control systems. We present also a high precise THz gas detection technique and its applications for noninvasive breath analysis and medical diagnostics.

2. Plasma Field Effect Transistor Arrays for Imaging in Sub-THz Atmospheric Windows

2.1. *Plasma field effect transistors THz detectors*

The possibility of the THz detection by FETs is due to nonlinearities of the plasma in the transistor. They lead to the rectification of the ac current induced by the incoming radiation. As a result, a photoresponse appears in the form of a dc voltage between source and drain which is proportional to the radiation intensity (photovoltaic effect). Depending on the frequency ω, one can distinguish two regimes of operation, and each of them can be further divided into two sub-regimes depending on the gate length. For the review see Refs. 1 and 2.

Figure 1 presents the schematics of FET as THz detector, where the incoming radiation creates an ac voltage with amplitude U_a only between the source and gate, U_0 is the static gate swing voltage ($U_0 = V_g - V_{th}$, where V_g is voltage between source and gate parts and V_{th} is the threshold voltage of transistor).

The FETs based THz detectors are very interesting for THz imaging because they operate at room temperature (in the overdamped plasma mode), being very fast, sensitive and having very high dynamic range.[5] They can also be used for helicity sensitive detection.[6]

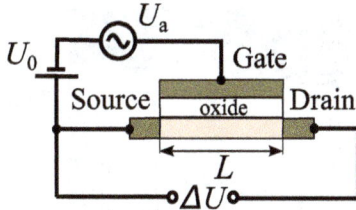

Fig. 1. Schematic of a FET operating as THz radiation detector. U_a represents ac voltage created by THz radiation between the source and gate, U_0 is the static gate swing voltage. ΔU is photovoltaic voltage, L is the length of transistor channel.

Recently, terahertz power dependence of the photoresponse of field effect transistors, operating at frequencies from 0.1 to 3 THz was investigated.[5] For Si metal–oxide–semiconductor field-effect transistors and InGaAs high electron mobility transistors the photoresponse increased linearly with increasing radiation intensity up to the ~10 kW/cm^2. Nonlinearity followed by saturation of the photoresponse was observed for higher intensities – see Fig. 2. The observed photoresponse nonlinearity was explained by the saturation of the transistor channel current. The theoretical model of terahertz field effect transistor photoresponse at high intensity was developed. The model explains quantitatively experimental data both in linear and nonlinear (saturation) range. Our results show that dynamic range of field effect transistors is very high and can extend over more than six orders of magnitudes of power densities (from ~ 0.5 mW/cm^2 to ~ 5 kW/cm^2).

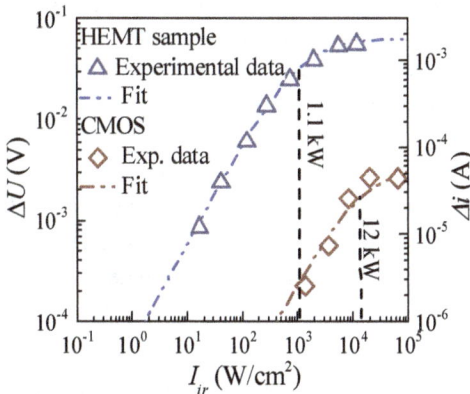

Fig. 2 Voltage and current generated by THz radiation. Saturation region is clearly visible. Dash-dot lines are results of calculations.[5]

The great advantage of THz-FET detectors is the possibility of their mass production and easy integration into arrays. In figure 3 we show an example of the imaging module based on Printed Circuit Boards (PCB's) technology. PCB's equipped with an array of 36 transistors and the parallel read-out electronic circuits are shown. On the left side of the Fig. 3 one can see the three PCB based panels of transistors integrated together to form three lines of ~ 250 mm length. These panels are equipped with an array of dedicated 3D printed diffractive lenses.

Fig. 3. THz modules forming matrix in postal scanner with additional matrix of lenses (a) and without (b).

2.2. *Diffractive 3D printed optics*

The first THz imaging systems were two axes raster scanning setups containing single point source and a single detector. They provided high quality images but the scanning time was relatively long, mainly limited by the speed of mechanical scanners. However, it appeared that to get the imaging speed acceptable for practical applications the linear detector arrays like these shown in Fig. 3 can be used. In the scanners, using linear detectors the object moves on the transportation belt and the image is constructed line by line using one or multiple linear detector arrays.

THz radiation sources present on the market are usually point sources. Therefore the active THz scanning using detector arrays require a special beam forming optics. 3D printing is a simple and cheap method of fabrication of THz optics. Recently, diffractive elements that can be 3D printed and that efficiently shape the THz beam into a line were demonstrated.[6,7] As shown in Ref. 7 by the use of a single diffractive optical element (hyperbolic–like lenses) one can convert a divergent THz beam into the focal line segment perpendicular to the optical axis. Such elements solve the problem of forming linear THz beams allowing together with adapted linear detector arrays (see Fig. 3) fast THz imaging in security and nondestructive quality control.

Experiments performed in the 300 GHz atmospheric window show that the printed multi-zone diffractive lens arrays offer both a good efficiency and good uniformity, and may improve the signal-to-noise ratio of the THz field effect transistor detector arrays (like these of Fig. 3) by more than one order of magnitude.

2.3. *Sub-THz imaging systems*

The arrays modules like these shown in Fig. 3, were applied in a fast on line imaging system (postal scanner) operating at 0.3 THz atmospheric window.

The control monitor of the postal scanner is shown in Fig. 4. It shows a real time image of the A4 envelope with CD disc in two viewing modes (THz-VIS fusion vision and pure THz image). The envelope is moving on the control belt.[7]

The THz FET imaging technology maybe used also for the control of welding process of glass fiber reinforced polymers in order to sort products according to conformity criteria. In this case an additional contrast may be used by analyzing the degree of circular polarization. Indeed, field effect transistor equipped with a special antenna may provide a detection voltage that depends on the degree of circular polarization of THz radiation.[4] First successful images with the additional polarization related contrast were obtained.[8]

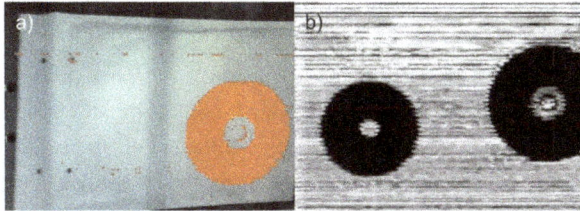

Fig. 4. The scan of CD discs carried out by postal mail scanner in two different viewing modes: fusion vision with THz image merged with the one in visible light (a) and only THz image of scanned envelope (b).

Fig. 5. The circular polarization contrast image of the glass fiber based aeronautic material.

In Fig. 5 one can see the circular polarization contrast image of the glass fiber based aeronautic material. The dark oblique areas show non-uniformity of the fibers distribution. This contrast was not visible using the amplitude based contrast imaging.

3. Sub-THz Spectroscopy for Health Protection

The great advantage of the THz range involves strong absorption lines (absorption coefficient can be $10^{-2} - 10^{-1}$ cm^{-1})[9] for many gases, including exhaled biomarkers (NO, CO, acetone, NH_3, etc.) that can ensure high sensitivity of spectroscopic methods. To get the high selectivity of analysis, the spectral resolution of about kHz is essential when providing detection, quantification, and monitoring of certain gases in multi-component mixtures, like exhaled breath. High resolution THz spectroscopy based on phase switching of radiation meets all the requirements for measurements at Doppler line resolution ($\sim 10^{-6}$) and frequency measurements with accuracy $10^{-8}-10^{-10}$.[10]

The periodic switching of the phase (with a π-shift) (or frequency) of radiation, interacting in resonance with gas, leads to the processes of transient radiation and absorption, that is periodic induction and decay of the macroscopic polarization.[11] The resulting transient signals are recorded and accumulated in the receiving part of the spectrometer. The value and shape of these signals are used for calculation of gas concentrations.

The maximum absorption coefficient which is proportional to the concentration of investigated gas is measured at central frequency of absorption line and then the concentration of gas marker is calculated. The dynamic of concentration of gas marker in multicomponent gas mixture can be investigated with using the change of absorption coefficient. Before exhaled breath measurements the gas cell is heated and evacuated to reduce a water vapor background. The breath samples are taken with a special bag, provided with an admission valve, thus preventing reverse air flow. The bag is attached to an evacuated gas cell via inlet valve. All measurements are carried out at working pressure of about 0.02 mbar and room temperature.

3.1. *Variation of the biomarkers concentration in diabetes patients' exhaled breath*

The investigations of acetone, methanol and ethanol concentration in the breath samples of 9 conditionally healthy volunteers and 8 diabetes patients (before and after taking the medicines) are carried out. For reference measurements breath samples of 9 healthy volunteers (6 men, 38 ± 10 years old; and 3 women, 35 ± 3 years old) were collected. The group of patients with type-2 diabetes comprised 8 men with a mean age of 63 ± 5 years was investigated. The precision of concentration determination was about 5%. An increase of acetone concentration in the diabetes patients breath in comparison with healthy volunteers: $(24,654 \div 35,235)$ ppm vs. $(0,904 \div 2,212)$ ppm correspondingly (Fig. 6) was found. Furthermore variation of the acetone concentration before and after taking the medicines is revealed. At the same time changing of the methanol and ethanol concentrations is found to be insignificant.

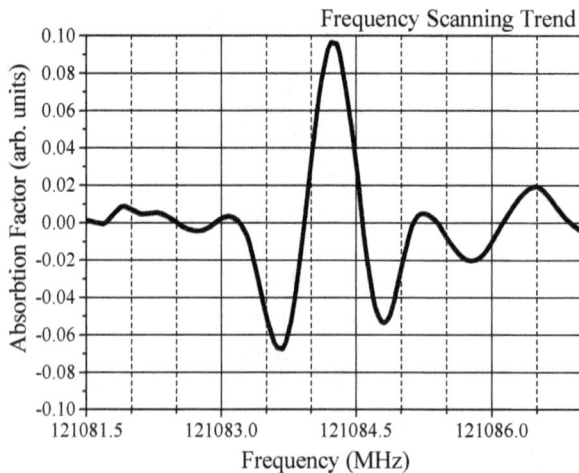

Fig. 6. The record of absorption line of acetone (in breath sample of diabetes patient) at the frequency 121.084 GHz.

Moreover, the absorption line of hydrogen sulfide (H_2S) at the frequency of 168.762 GHz is detected in the exhaled breath of one diabetes patient. The H_2S concentration in the exhaled breath of diabetes patient is increased about 4 times in comparison with that one in exhaled breath of a healthy volunteer.

3.2. *NO concentration in breath at the radiotherapy course*

The detection of NO is performed at a frequency of 150.176 GHz. The measurements were carried out on the base of radiology ward of Nizhny Novgorod regional oncology centre. The study is done with 5 healthy volunteers and various groups of patients with stage 3 and 4 lung cancer (5 men, 59 ± 10 years old; and 1 woman, 50 years old) and stage 2 Hodgkin's lymphoma (2 females, 48 and 19 years old). The measurement results have demonstrated a 2–3 times increased NO concentration in exhaled breath of lung cancer patients after a session of radiation therapy. The linear accelerator Philips SL in deceleration emission mode (with energy of 5 MeV) was used for radiotherapy. An initial tumor and area of healthy lung tissue around it and also lymphatic glandes of mediastinum was included in radiation treatment for patient with lung cancer. The summary dose to investigation time was 12 gray. The probes of exhaled breath were carried out during 10 s before and 10 s after radiation treatment. The experimental results of NO concentration measurements are summarized in Table 1.

Table 1. Results of NO concentration measurements.

Patient	NO concentration before radiotherapy, ppm	Level
Healthy volunteers (n=5)	33÷75	-
Stage 2 Hodgkin's lymphoma (n=2)	120÷149	220÷254
Stage 3 central endobronchial cancer of lung (n=5)	149÷254	335÷750
Stage 4 central endobronchial cancer of lung (n=1)	1200	2240

3.3. *NH₃ concentration in breath at gastrointestinal tract diseases*

The gas analyzer has also been employed for detection of exhaled NH_3 in breath samples of patients with *Helicobacter pylori* infection (3 men, 25 ± 5 years old; and 2 women, 30 and 27 years old). Ammonia concentration in patients' breath is found to be one order higher than in the breath of healthy volunteers, (75,125÷100,415) ppm vs. (8,752÷10,453) ppm correspondingly.

The high precise THz spectrometer meets all requirements for exhaled breath diagnostics. Although, its sensitivity is less than that of the best chromatograph-mass spectrometers, the THz spectrometer ensures other advantages. The most important one is unique and reliable identification of a gas marker in contrast to chromatograph-mass spectrometers which can give only identification probability. Furthermore, the THz spectrometer is not expensive, easy-to-use and doesn't require high-priced components and consumables.

4. Conclusion

Recent developments of THz detector arrays based on plasma wave field effect transistors were demonstrated. By simultaneous development of the transistor arrays with their read-out circuits on PCB electronic boards and a special diffractive 3D printed optics we demonstrated systems for imaging in 300 GHz atmospheric window based on fast linear scanner. The first high speed THz postal scanner developed for real time fast A4 envelopes security screening is presented.

Also in the present work were demonstrated capabilities of the high precise THz spectrometry for analysis of exhaled breath as a multicomponent gas mixture. The gas analyzers, based on the THz spectrometers, have been successfully applied for detection of the important biomarkers (NO, acetone, alcohols) in exhaled breath of various patients and healthy people. The approach, presented in this paper, could be promising to become the optimal "electronic nose" for exhaled breath diagnostics.

Acknowledgments

This work was financially supported by the National Centre for Research and Development in Poland (grant no. PBS1/A9/11/2012), by the National Science Centre in Poland (DEC-2013/10/M/ST3/00705) by CNRS France via LIA –TERAMIR projects and by COST –MP1204 TERAMIR action and by ERA.Net RUS plus Program (RFBR Grant No 16-52-76011). The authors would like to thank the Orteh Company for providing LS 6.0 Software for designing and modeling of the diffractive optical elements and for delivering THz matrices for tests.

References

1. W. Knap and M. Dyakonov, "Field effect transistors for terahertz applications", in *Handbook of Terahertz Technology*, ed. D. Saeedkia, Woodhead Publishing, Waterloo, Canada, 2013, pp. 121–155.
2. W. Knap, S. Rumyantsev, M. Vitiello, D. Coquillat, S. Blin, N. Dyakonova, M. Shur, F. Teppe, A. Tredicucci, and T. Nagatsuma, "Nanometer size field effect transistors for terahertz detectors", *Nanotech.* **24(21)** (2013) 214002.
3. O. A. Klimenko, W. Knap, B. Iniguez, D. Coquillat, Y. A. Mityagin, F. Teppe, N. Dyakonova, H. Videlier, D. But, F. Lime, J. Marczewski, and K. Kucharski, "Temperature enhancement of terahertz responsivity of plasma field effect transistors", *J. Appl. Phys.* **112(1)** (2012) 014506.
4. C. Drexler, N. Dyakonova, P. Olbrich, J. Karch, M. Schafberger, K. Karpierz, Y. Mityagin, M. B. Lifshits, F. Teppe, O. Klimenko, Y. M. Meziani, W. Knap, and S. D. Ganichev, "Helicity sensitive terahertz radiation detection by field effect transistors", *J. Appl. Phys.* **111(12)** (2012) 124504.
5. D. B. But, C. Drexler, M. V. Sakhno, N. Dyakonova, O. Drachenko, F. F. Sizov, A. Gutin, S. D. Ganichev, and W. Knap, "Nonlinear photoresponse of field effect transistors terahertz detectors at high irradiation intensities", *J. Appl. Phys.* **115(16)** (2014) 164514.
6. J. Suszek, A. Siemion, M. S. Bieda, N. Blocki, D. Coquillat, G. Cywinski, E. Czerwinska, M. Doch, A. Kowalczyk, N. Palka, A. Sobczyk, P. Zagrajek, M. Zaremba, A. Kolodziejczyk, W. Knap, and M. Sypek, "3-D-printed flat optics for THz linear scanners", *IEEE Trans. Terahertz Sci. Technol.* **5(2)** (2015) 314–316.

7. Mail scanner: http://www.unipress.waw.pl/fastthzscanner/ and http://www.orteh.pl/page/22/research-development
8. 2D Imaging system: http://www.t-waves-technologies.com/en
9. H. M. Pickett, E. A. Cohen, B. J. Drouin, J. C. Pearson *et al.*, "Submillimeter, Millimeter, and Microwave Spectral Line Catalog", *JPL Molecular Spectroscopy*, California Institute of Technology (http://spec.jpl.nasa.gov/ftp/pub/catalog/catform.html).
10. V. Vaks, "High-precise spectrometry of the terahertz frequency range: The methods, approaches and applications", *J. Infrared Millim. Terahertz Waves* **33(1)** (2012) 43–53.
11. V. L. Vaks, E. G. Domracheva, E. A. Sobakinskaya, and M. B. Chernyaeva, "Exhaled breath analysis: physical methods, instruments, and medical diagnostics", *Physics – Uspekhi* **57(7)** (2014) 684–701.

Calculation of Modal Gain for Terahertz Lasers Based on HgCdTe Heterostructures with Quantum Wells

Alexander A. Dubinov

Institute for Physics of Microstructures of Russian Academy of Sciences,
Nizhny Novgorod 603950, Russia
sanya@ipmras.ru

Vladimir Ya. Aleshkin

Institute for Physics of Microstructures of Russian Academy of Sciences,
Nizhny Novgorod 603950, Russia

In this work we calculate and analyze modal gain for terahertz lasers based on HgCdTe heterostructures with quantum wells (QWs) taking into account the symmetry-enforced light hole-heavy hole mixing at the quantum well interfaces. We have found that modal gain for a structure with 5 HgTe QWs of the 5.2 nm width can be 33 cm^{-1} at 9 THz.

Keywords: Gain; quantum well; terahertz frequency.

1. Introduction

In recent years a considerable effort was put in furthering the infrared spectroscopy to terahertz frequency range, since this spectral region includes the characteristic absorption bands of many organic and inorganic molecules. Compact tunable emitters are in demand for infrared spectroscopy. Quantum cascade lasers (QCL) are successfully used for spectroscopy in THz range.[1] However, complex design of QCL structures is highly demanding and requires an elaborate technology that was only realized on A3B5 semiconductors. This results in formidable problem of developing QCL for 6–15 THz since this spectral region is characterized by strong phonon absorption in A3B5 materials.

On the other hand, ternary semiconductor compounds with heavier elements, in particular HgCdTe have lower phonon frequencies (3–5 THz), variable bandgap energy covering the wide spectral range 0–1.6 eV. This allows the radiation emission to terahertz frequency range by interband optical transitions at non-equilibrium pumping. Previously, lasing in HgCdTe structures has been studied at mid infrared frequency range (55–150 THz).[2]

Recent investigations into properties of HgCdTe epitaxial structures that are grown at low temperatures by the molecular-beam epitaxy reveal the remarkable quality of such structures. Thus, narrow gap HgCdTe epilayers demonstrate pronounced photosensitivity in terahertz range,[3] photoluminescence (PL) above 11 THz[4] and stimulated emission (SE) at 35 THz under an optical pump.[5] Finally, Morozov *et al.*[6] demonstrated SE from HgCdTe waveguide structures with QWs above 31 THz with a low threshold of 0.12 kW/cm^2. Kinetics of PL in QW structures show that the carrier concentration necessary for light amplification can be achieved under reasonable pumping intensity.[7]

The possibility of optical gain in HgCdTe QWs in 5–25 THz range has been already theoretically predicted.[8] According to the performed calculations with the use of the 8×8 Kane Hamiltonian[9] for 5.6 nm (close to the critical thickness, when Dirac cone exists[10]) HgTe QW in Hg$_{0.3}$Cd$_{0.7}$Te barriers, both the minimum in the conduction band and the maximum in the valence band for this structure are arranged at point $k = 0$ and the band gap E_g is 20 meV, which corresponds to a frequency of 4.8 THz.

However, Tarasenko *et al.*[11] described the fine structure of Dirac states in the HgTe/CdTe quantum wells of critical and close-to-critical thicknesses and shown that the necessary creation of interfaces brings in another important physical effect: the opening of a significant anticrossing gap between the tips of the Dirac cones. They presented a microscopic theory of the Dirac states in HgTe/CdTe QWs that predicts a very large [15 meV (3.6 THz)] anticrossing gap between the tips of the Dirac cones in QWs of a critical thickness by symmetry-enforced light hole-heavy hole mixing at the quantum well interfaces.

In this work we calculate and analyze modal gain for terahertz lasers based on HgCdTe heterostructures with quantum wells taking into account the above-mentioned effects.

2. Model for Calculation of Matrix Element of the Coordinate Operator

Kane 8×8 model was used for calculations. The Hamiltonian was selected as used in Ref. 12 with the addition of the term of the formula (2) from Ref. 11. Electron wave functions were calculated with the method described in Ref. 9. The essence of this method is to consider a superlattice with sufficiently long period d instead of a single QW. Tunneling of electrons in the superlattice can be neglected on account of the large thickness of the barriers between the QWs. Thus, one can use the condition of the wave function periodicity (with a period d) as the boundary condition. Electronic wave function has 8 components in this model. Each component can be decomposed in spatial harmonics:

$$\Psi_j(z, y, z) = \exp(ik_x x + ik_y y) \sum_{n=-N_{max}}^{N_{max}} C_n(k_x, k_y) \exp(i\frac{2\pi z}{d}n) \quad (1)$$
$$= \exp(ik_x x + ik_y y)\varphi_j(k_x, k_y, z),$$

where $k_{x,y}$ are the electron wave vector components in QW, z is the normal coordinate to the QW plane, $2N_{max} + 1$ is the number of the spatial harmonics in the expansion, index j denotes the component of the wave function, $C_n(k_x, k_y)$ are the expansion coefficients. We assume that the quantum wells have been grown on a (013) plane and choose $x \parallel [100]$, $y \parallel [03\bar{1}]$, $z \parallel [013]$, because high quality structures are grown on semi-insulating GaAs(013) substrates with ZnTe and CdTe buffers usually.[13] In the framework of this approach finding the electron spectrum and wave functions reduces to finding the eigenvalues and eigenvectors of the matrix with $8(2N_{max}+1) \times 8(2N_{max}+1)$ dimension. We used $N_{max} = 40$ in our calculation. The value of d was chosen to be 30 nm.

Figure 1 shows the dependence of the electron energy on the wave vector k_x for the size quantized subbands of the HgTe/Cd$_{0.7}$Hg$_{0.3}$Te structure with HgTe QW of the 6.3 nm width at temperature T = 4.2 K and for two cases: taking and not taking into account light hole-heavy hole mixing at the quantum well interfaces. From Fig. 1 one can see, that the energy spectrum strong changes around point $k_x = 0$ for taking into account light hole-heavy hole mixing: the band gap [$\sim$ 11 meV (2.6 THz)] is formed and splitting of subbands occurs. For a case HgTe/Cd$_{0.7}$Hg$_{0.3}$Te structure with HgTe QW of the 5.2 nm width the band gap E_g increases to 37 meV (9 THz).

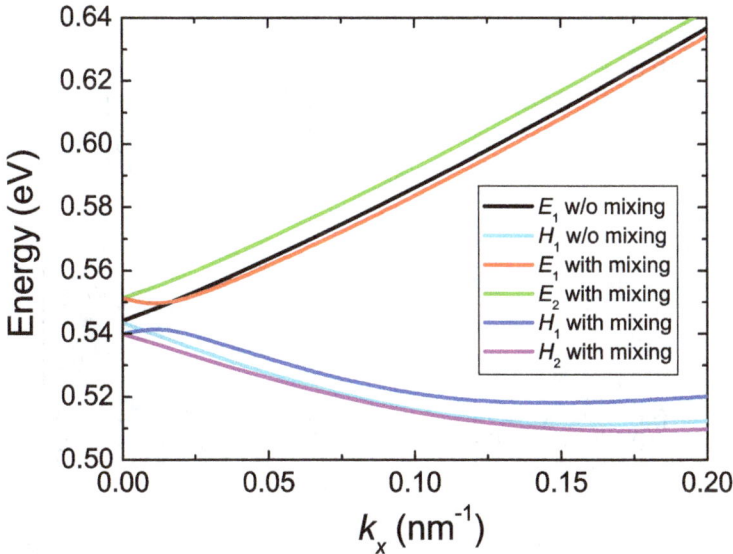

Fig. 1. The energy spectrum for the HgTe/Cd$_{0.7}$Hg$_{0.3}$Te (013) structure with the 6.3 nm width HgTe QW at T = 4.2 K from the calculation taking and not taking into account light hole-heavy hole mixing. E_1 and E_2 are the conduction subbands, H_1 and H_2 are the valence subbands.

The matrix element of x, y, z-coordinate operator is a convenient parameter to characterize the electron transitions between the initial state $|i>$ and the final state $|f>$ via absorption of light polarized along the x, y, z-axis, correspondently. For example, matrix element of x-coordinate operator is

$$x_{fi} = -i\hbar \frac{\dot{x}_{fi}}{\varepsilon_f - \varepsilon_i}, \tag{2}$$

where ε_i and ε_f are the electron energies in the initial and final states, $\hbar$ is the Planck constant, $\dot{x}_{fi}$ is the matrix element of the x-component of velocity operator. The expression for the velocity operator was obtained from the relation

$$\dot{x}_{fi} = \frac{i}{\hbar} (Hx - xH), \tag{3}$$

where H is the Hamiltonian of a system.

Figure 2 shows the calculated modulus of the matrix elements (between the states of the electron and hole subbands) for the x-coordinate operator of structure with HgTe QW of the 5.2 nm width. One can see that matrix elements x_{E1H1} and x_{E2H2} are nonzero and have maximum at $k_x \neq 0$, but x_{E1H2} and x_{E2H1} are nonzero at $k_x \approx 0$ only.

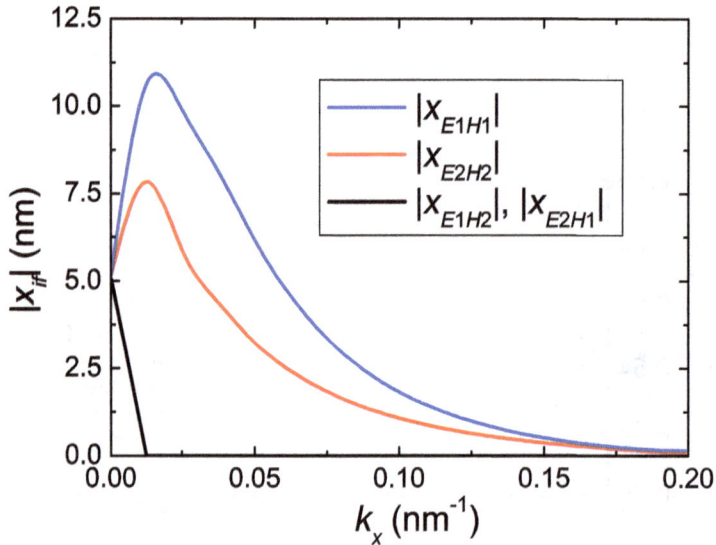

Fig. 2. Dependences of the modulus of the matrix elements $|x_{if}|$ on the wave vector.

Since the splitting is rather small (~ 5 meV for HgTe QW of the 5.2 nm width), it is convenient to characterize interband transitions by the following values:

$$x_{EH} = \sqrt{|x_{E1H1}|^2 + |x_{E1H2}|^2 + |x_{E2H1}|^2 + |x_{E2H2}|^2}, \tag{4}$$

because the summary transition probability between electron and hole subbands is proportional to $|x_{EH}|^2$.

3. Calculation of Surface Conductivity and Modal Gain

The real part of the surface conductivity $\text{Re}(\sigma)$ of a non-equilibrium electron-hole system, which determines the absorption or gain coefficient of photons with the frequency ω, comprises the contributions of both the interband and intraband transitions:

$$\text{Re}(\sigma) = \text{Re}(\sigma^{inter}) + \text{Re}(\sigma^{intra}). \tag{5}$$

For a parabolic approximation of energy spectrum (for non-equilibrium carrier concentration less than 2×10^{10} cm^{-2} in the HgTe QW of the 5.2 nm width) expression for the interband surface conductivity will be the following:

$$\text{Re}(\sigma^{inter}) = \frac{e^2 \omega m^*}{2\hbar^2} |x_{EH}|^2 \left[\Theta(\hbar\omega - E_g - E_F) - \Theta(\hbar\omega - E_g) \right], \tag{6}$$

where e is the electron charge, m^* is the electron mass (0.0038 from free electron mass in our case), E_F is the quasi-Fermi energy (11 meV for considered concentration).

For intraband surface conductivity one can use the Drude formula:

$$\text{Re}(\sigma^{intra}) = \frac{2e\mu n}{1 + \omega^2 \tau^2}. \tag{7}$$

Here, μ is the electron and hole mobility in HgTe QW, τ is the momentum relaxation time of electrons and holes associated with their scattering, and n is the electron and hole concentration. The electron and hole mobility and momentum relaxation time were assumed to be $\mu = 10^5$ cm^2/(Vs) and $\tau = 65$ ps,[8] respectively.

Figure 3 shows the frequency dependence of the real part of the surface conductivity. From Fig. 3 one can see, that $\text{Re}(\sigma)$ have the maximum at frequency 9 THz, which corresponds to the band gap E_g for this structure. The value of the maximum of $\text{Re}(\sigma)$ for HgTe QW is comparable with the value of $\text{Re}(\sigma)$ for graphene.[14]

Let's consider the structure HgTe/Cd$_{0.7}$Hg$_{0.3}$Te shown in Fig. 4. The thickness of ZnTe, CdTe and Cd$_{0.7}$Hg$_{0.3}$Te layers are 50 nm, 10 μm and 6 μm, correspondingly. The structures were designed so as to effectively confine light to in-plane direction, i.e., provide a waveguide for radiation with the frequency corresponding to the bandgap energy. The electrodynamic calculations of the TE mode localization were performed by the transfer matrix method. Data on the refractive index were taken from the data handbook.[15] Figure 4 illustrates the TE$_0$ mode localization in the structure also. As can be seen from Fig. 4, the active region consists five QWs that were grown inside the dielectric waveguide layer. The active region was centered on the antinode of the fundamental TE$_0$ mode of the waveguide. This was done to maximize the overlap between the photons and electronic excitations.

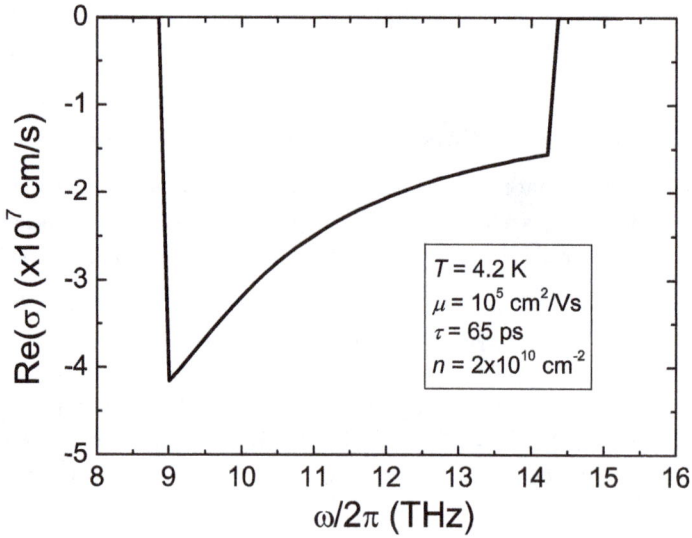

Fig. 3. The frequency dependence of the real part of the surface conductivity.

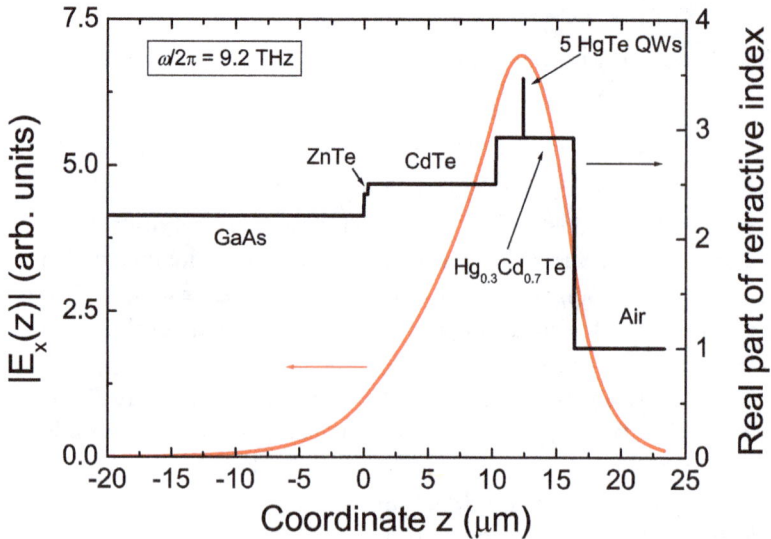

Fig. 4. The coordinate dependences of the real part of refractive index and the modulus of x-component of electric field in TE$_0$ mode.

The modal gain, which describes the attenuation or amplification of the propagating mode, can be calculated using the following formula:

$$g = -\frac{4\pi \mathrm{Re}(\sigma)}{c n_{eff}} \Gamma - \alpha, \qquad (8)$$

where c is the light speed in vacuum, n_{eff} and α are the effective refractive index and the absorption of TE_0 mode, correspondingly, Γ is the gain-overlap factor:

$$\Gamma = \frac{\sum_{j=1}^{5} |E_x(z_j)|^2}{\int_{-\infty}^{+\infty} |E_x(z)|^2 dz}. \tag{9}$$

Here z_j are the coordinates of HgTe QWs in the structure. Figure 5 shows the frequency dependences of n_{eff}, α and Γ. From Fig. 5 one can see that α decreases but n_{eff} and Γ increase with frequency growth. Also there are local maximum of α and local minimum of Γ at 10 THz, which are associated with the strong Reststrahlen band absorption in GaAs substrate.

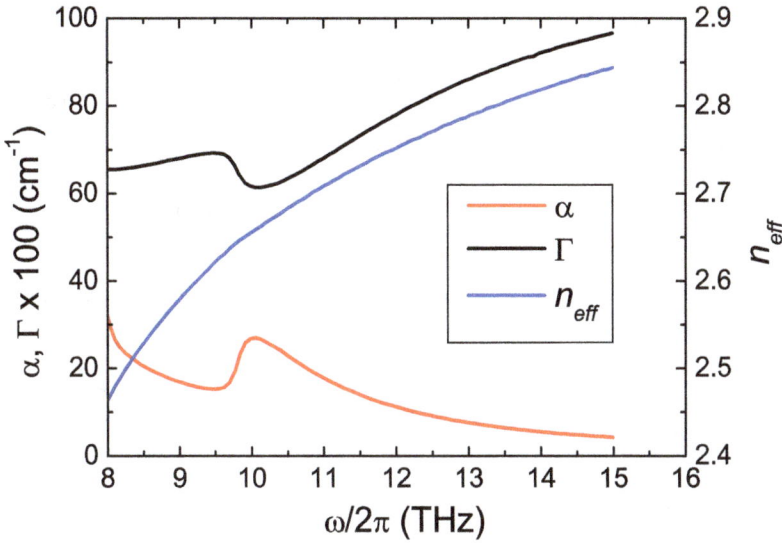

Fig. 5. The frequency dependences of n_{eff}, α and Γ.

Figure 6 shows the frequency dependence of the modal gain g. As can be seen from Fig. 6, there are two maximums of g at 9 and 14 THz, respectively. The reason of this effect is a general behavior of $\text{Re}(\sigma)$, n_{eff}, α and Γ on frequency. The value of maximum of g is 33 cm^{-1} at 9 THz, which is comparable with the value of g for THz QCLs.[16]

4. Conclusion

We have calculated matrix elements for the electron - hole interband transitions in HgTe/Cd$_{0.7}$Hg$_{0.3}$Te structure with HgTe QW using 8×8 Kane model that takes into account the symmetry-enforced light hole-heavy hole mixing at the quantum well interfaces. We have shown that real part of the surface conductivity of a non-equilibrium electron-hole system in this structure can be comparable with the value

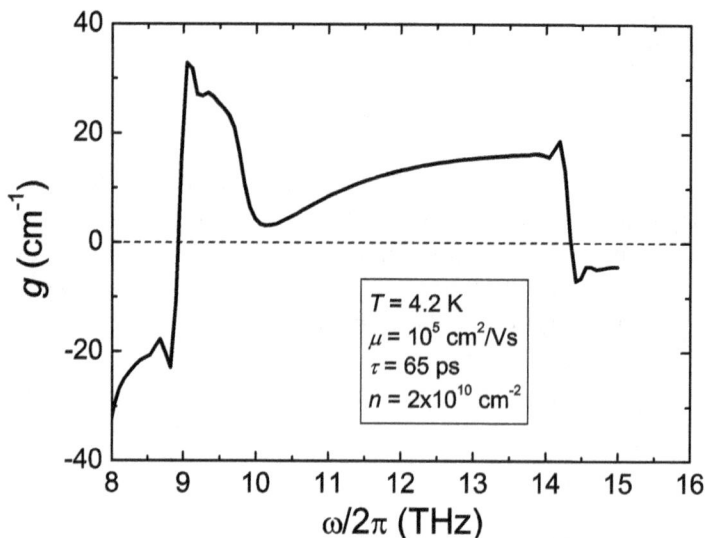

Fig. 6. The frequency dependence of modal gain g.

of real part of the surface conductivity for graphene. We have found that modal gain at 9 THz for a structure with 5 HgTe QWs of the 5.2 nm width can be comparable with the value of the modal gain for THz QCLs.

Acknowledgments

This work was financially supported by Russian Foundation for Basic Research (#15-02-08274, #15-42-02249), Ministry of Education and Science of the Russian Federation (President Program of Support for Leading Scientific Schools of Russia #7577.2016.2) and the Russian Academy of Sciences.

References

1. M. S. Vitiello, G. Scalari, B. Williams, and P. De Natale, "Quantum cascade lasers: 20 years of challenges", *Opt. Exp.* **23** (2015) 5167–5182.
2. J. M. Arias, M. Zandian, R. Zucca, and J. Singh, "HgCdTe infrared diode lasers grown by MBE", *Semicond. Sci. Technol.* **8** (1993) S255–S260.
3. V. V. Rumyantsev, S. V. Morozov, A. V. Antonov, M. S. Zholudev, K. E. Kudryavtsev, V. I. Gavrilenko, S. A. Dvoretskii, and N. N. Mikhailov, "Spectra and kinetics of THz photoconductivity in narrow-gap $Hg_{1-x}Cd_xTe$ $(x < 0.2)$ epitaxial films", *Semicond. Sci. Technol.* **28** (2013) 125007.
4. S. V. Morozov, V. V. Rumyantsev, A. V. Antonov, K. V. Maremyanin, K. E. Kudryavtsev, L. V. Krasilnikova, N. N. Mikhailov, S. A. Dvoretskii, and V. I. Gavrilenko, "Efficient long wavelength interband photoluminescence from HgCdTe epitaxial films at wavelengths up to 26 μm", *Appl. Phys. Lett.* **104** (2014) 072102.
5. S. V. Morozov, V. V. Rumyantsev, A. A. Dubinov, A. V. Antonov, A. M. Kadykov, K. E. Kudryavtsev, D. I. Kuritsin, N. N. Mikhailov, S. A. Dvoretskii, and

V. I. Gavrilenko, "Long wavelength superluminescence from narrow gap HgCdTe epilayer at 100 K", *Appl. Phys. Lett.* **107** (2015) 042105.

6. S. V. Morozov, V. V. Rumyantsev, A. M. Kadykov, A. A. Dubinov, K. E. Kudryavtsev, A. V. Antonov, N. N. Mikhailov, S. A. Dvoretskii, and V. I. Gavrilenko, "Long wavelength stimulated emission up to 9.5 μm from HgCdTe quantum well heterostructures", *Appl. Phys. Lett.* **108** (2016) 092104.

7. S. V. Morozov, V. V. Rumyantsev, A. V. Antonov, A. M. Kadykov, K. V. Maremyanin, K. E. Kudryavtsev, N. N. Mikhailov, S. A. Dvoretskii, and V. I. Gavrilenko, "Time resolved photoluminescence spectroscopy of narrow gap $Hg_{1-x}Cd_xTe/Cd_yHg_{1-y}Te$ quantum well heterostructures", *Appl. Phys. Lett.* **105** (2014) 022102.

8. S. V. Morozov, M. S. Joludev, A. V. Antonov, V. V. Rumyantsev, V. I. Gavrilenko, V. Ya. Aleshkin, A. A. Dubinov, N. N. Mikhailov, S. A. Dvoretskiy, O. Drachenko, S. Winnerl, H. Schneider, and M. Helm, "Study of lifetimes and photoconductivity relaxation in heterostructures with $Hg_xCd_{1-x}Te/Cd_yHg_{1-y}Te$ quantum wells", *Semiconductors* **46** (2012) 1362–1366.

9. A. V. Ikonnikov, M. S. Zholudev, K. E. Spirin, A. A. Lastovkin, K. V. Maremyanin, V. Ya. Aleshkin, V. I. Gavrilenko, O. Drachenko, M. Helm, J. Wosnitza, M. Goiran, N. N. Mikhailov, S. A. Dvoretskii, F. Teppe, N. Diakonova, C. Consejo, B. Chenaud, and W. Knap, "Cyclotron resonance and interband optical transitions in HgTe/CdTe (013) quantum well heterostructures", *Semicond. Sci. Technol.* **26** (2011) 125011.

10. B. Buttner, C. X. Liu, G. Tkachov, E. G. Novik, C. Brune, H. Buhmann, E. M. Hankiewicz, P. Recher, B. Trauzettel, S. C. Zhang, and L. W. Molenkamp, "Single valley Dirac fermions in zero-gap HgTe quantum wells", *Nature Physics* **7** (2011) 418–422.

11. S. A. Tarasenko, M. V. Durnev, M. O. Nestoklon, E. L. Ivchenko, J.-W. Luo, and A. Zunger, "Split Dirac cones in HgTe/CdTe quantum wells due to symmetry-enforced level anticrossing at interfaces", *Phys. Rev. B* **91** (2015) 081302(R).

12. E. G. Novik, A. Pfeuffer-Jeschke, T. Jungwirth, V. Latussek, C. R. Becker, G. Landwehr, H. Buhmann, and L. W. Molenkamp, "Band structure of semimagnetic $Hg_{1-y}Mn_yTe$ quantum wells", *Phys. Rev. B* **72** (2005) 035321.

13. V. S. Varavin, V. V. Vasiliev, S. A. Dvoretsky, N. N. Mikhailov, V. N. Ovsyuk, Y. G. Sidorov, A. O. Suslyakov, M. V. Yakushev, and A. L. Aseev, "HgCdTe epilayers on GaAs: growth and devices", *Proc. SPIE* **5136** (2003) 381–395.

14. V. Ya. Aleshkin, A. A. Dubinov, and V. Ryzhii, "Terahertz laser based on optically pumped graphene: model and feasibility of realization", *JETP Lett.* **89** (2009) 63–67.

15. O. Madelung, *Semiconductors: Data Handbook*, Springer, New York, 2003.

16. D. Indjin, P. Harrison, R. W. Kelsall, and Z. Ikonic, "Self-consistent scattering model of carrier dynamics in GaAs-AlGaAs terahertz quantum-cascade lasers", *IEEE Photon. Technol. Lett.*, **15** (2003) 15–17.

Plasmonic Enhancement of Terahertz Devices Efficiency

Vladimir Mitin

Department of Electrical Engineering, University at Buffalo,
SUNY, Buffalo, NY 14260-1920, USA
vmitin@buffalo.edu

Victor Ryzhii

Research Institute of Electrical Communication, Tohoku University,
Sendai 980-8577, Japan
Institute of Ultra High Frequency Semiconductor Electronics of RAS,
Moscow 117105, Russia

Maxim Ryzhii

Department of Computer Science and Engineering, University of Aizu,
Aizu-Wakamatsu 965-8580, Japan

Akira Satou and Taiichi Otsuji

Research Institute of Electrical Communication, Tohoku University,
Sendai 980-8577, Japan

Michael S. Shur

Department of Electrical, Computer, and Systems Eng., Rensselaer Polytechnic Institute,
Troy, NY 12180, USA

This paper reviews the plasmonic effects in graphene THz photodetectors (PD) and light emitters (LE). It is demonstrated that the devices based on double graphene-layer (DGL) or multiple graphene-layer structures with the graphene layers separated by thin tunnel barrier layers have advantages over the single graphene-layer (SGL) devices. In DGLs, this advantage is due to the photon-assisted resonant tunneling when the band offset of the graphene layers is aligned to the THz photon energy. The resonant emission or absorption of the THz radiation is enhanced by the cooperative resonant excitation of the graphene plasmons leading to an extremely high gain and/or responsivity in the graphene THz device structures.

Keywords: Graphene devices; double graphene-layer devices; THz photodetectors; THz lasers/ emitters.

1. Introduction

Plasmonic electron phenomena in semiconductors could enable advanced terahertz (THz) devices.[1,2] In particular, the THz detectors using the non-linear plasmonic properties of two-dimensional electron gas (2DEGs) have proved to be very effective devices operating at room temperature.[3-5] Unique properties of 2D electrons and 2D holes in graphene make SGL and DGL structures to be especially promising for the THz devices. As described below, these devices have already demonstrated superior plasmonic enhancement at room temperature.

2. Results and Discussion

The p-i-n structures using SGLs and multiple graphene layer structures can enable efficient photodetectors in the THz and IR frequency ranges (see for example Ref. 6 and review Ref. 7). However, such p-i-n diodes practically have no selectivity since the coefficient of absorption of graphene does not depend on frequency. This restricts their usage in spite of their good performance. It was shown[8] that the plasma oscillations in graphene enhance the detection responsivity at the resonant frequencies and, therefore, introduces photodetection selectivity. Analogous plasmonic enhancement is observed in emission of THz radiation by graphene when frequency of surface plasmon polaritons is close to the frequency of negative dynamic conductivity under the optical pumping[9] (an example is shown in Fig. 1 where gain for lasing structure is reaching 10^4 cm^{-1}).

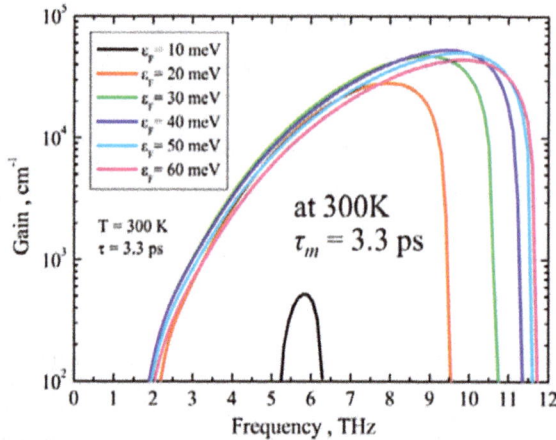

Fig. 1. Dependence of gain on frequency for a monolayer population-inverted graphene on SiO$_2$/Si substrate for different quasi-Fermi energies that are shown on the insert of the figure. Carrier momentum relaxation time 3.3 ps and temperature 300 K are also shown in the figure.[9]

The DGL heterostructures with a thin inter-GL barrier (like h-BN, WS$_2$, and others) have a substantial advantage over a SGL enabled by photon-assisted resonant radiative inter-GL transitions[10-12] as it is demonstrated in Fig. 2. This brings selectivity to PDs and LEs. The main element of both the LEs and PDs under consideration is a DGL core-shell heterostructure with the independently contacted GLs separated by a thin transparent tunnel-barrier layer. The bias voltage V applied between the GL contacts induces the electron and hole gases in the opposing GLs controlled by the gate voltage V_g as shown in Fig. 3. The outer gate-stack structures at both sides of the DGL core shell also serve as the metal-metal (MM) waveguides. The voltage-dependent band-offset energy between the Dirac points of the GLs (designated as 'Δ' in Fig. 2) and the depolarization shift determine the energies of the photons emitted (in the LEs) or absorbed (in the PDs) due to the resonant-tunneling inter-GL transitions shown in Fig. 2. The outer gate electrode can be configured in a grating-gate structure serving as a broadband antenna[12] as shown in Fig. 3.

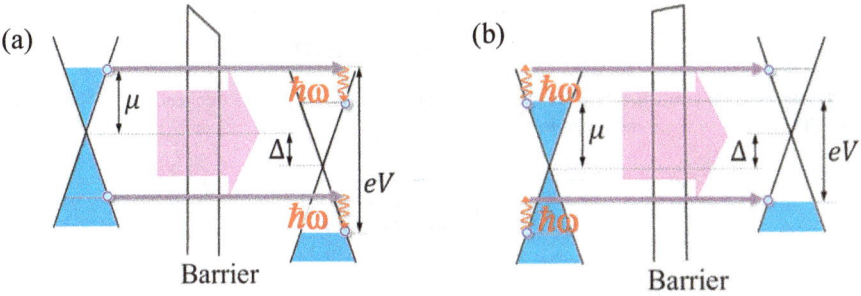

Fig. 2. Band diagrams of (a) photoemission-assisted inter-GL radiative transition and (b) photo-absorption-assisted inter-GL transitions.

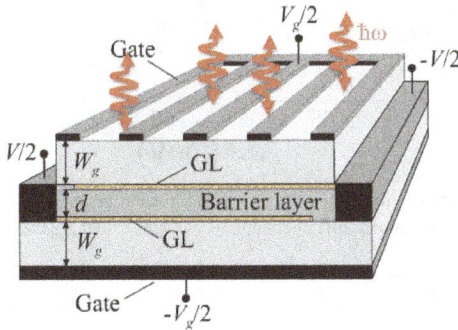

Fig. 3. A DGL heterostructure for THz LEs and PDs.[12]

Figure 4 shows an example of the plasmonic enhancement of the PD responsivity in the DGL structure with different values of the electron and hole collision frequencies that are indicated in the figure for the plasma frequency Ω of 3 THz. There are several resonant peaks associated with plasma oscillations. The highest peaks correspond to the zero resonant frequency $\omega_0 < \Omega$ and the subsequent resonances correspond to the multiples of the plasma frequency. Plasmonics promotes the double resonance and the plasmonic enhancement can reach several orders of magnitude compared with just a single resonance (see Fig. 4). Figure 5 demonstrates a very effective tunability of the response resonant frequency by the applied bias V in the DGL structure[12] shown in Fig. 3.

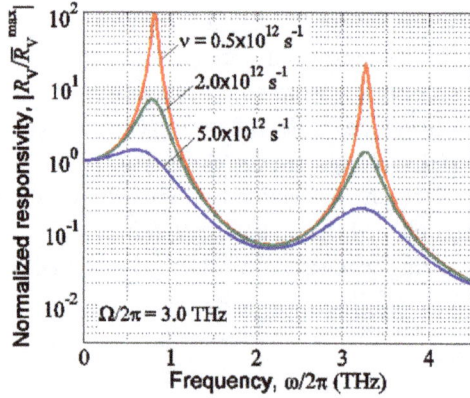

Fig. 4. Dependence of normalized responsivity on signal frequency for plasma frequency 3.0 THz and different collision frequencies of electrons and holes.[11]

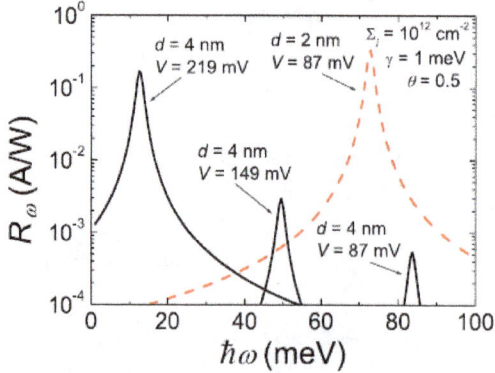

Fig. 5 Simulated DGL-PD responsivity vs. photon energy $\hbar\omega$ for the inter-GL barrier layer thickness $d = 4$ nm (solid lines) and $d = 2$ nm (dashed line) at different applied voltages V that are indicated in the figure.[13]

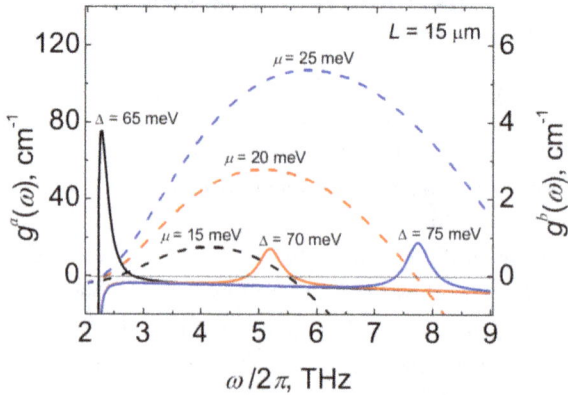

Fig. 6. Simulated frequency dependence of the THz gain for the DGL inter-GL transition laser (solid lines and left axis) for different band-offset energies between the Dirac points Δ and of the THz gain for the DGL intra-GL transition laser (dashed lines and right axis) for different Fermi energies that are indicated at the curves.[10]

Figure 6 demonstrates more than 10 times enhancement of the gain due to the resonant tunneling between the GLs in the DGL structure[11] shown in Fig. 3 enhanced by the plasmons compared to the case of the same two GL structure with conventional intraband transitions in each GL. The length and the width of GLs are 15 μm and 5 μm, respectively. The resonant peek in the case of the two resonances is very narrow.

Recently we demonstrated[13] that the plasmonic effect in a GL serving as a base in the vertical graphene-base hot-electron transistors (GB-HETs) leads to the possibility of the GB-HET operation at the frequencies significantly exceeding those limited by the characteristic RC-time. It was found that the responsivity of GB-HETs with a sufficiently perfect GB exhibits sharp resonant maxima at the THz frequencies associated with the excitation of plasma oscillations in the GL base. The positions of these maxima are controlled by the applied bias voltages. These GB-HETs surpass other plasmonic THz detectors.

Plasmonic effects are strongly pronounced in PDs[14] based on field-effect transistors (FETs) with split gates, electrically induced lateral p–n junctions, uniform GL [Fig. 7(a)] or perforated (in the p–n junction depletion region) graphene layer (PGL) [Fig. 7(b)) channel. The perforated depletion region forms an array of the nano-constrictions or nanoribbons creating the barriers for the holes and electrons as shown in Figs. 7(c) and 7(d). The operation of the GL-FET- and PGL-FET-detectors is associated with the rectification of the ac current across the lateral p–n junction enhanced by the excitation of bound plasmonic oscillations in the p- and n-sections of the channel. These detectors can exhibit very high voltage responsivity at the THz radiation frequencies close to the frequencies of the plasmonic resonances. Due to much lower p–n junction conductance in the PGL-FET-detectors, their resonant response can be substantially more pronounced than in the GL-FET-detectors corresponding to fairly high detector responsivity. An example is shown in Fig. 8 for PGL-FET for different Fermi energies and carrier collision frequency $0.5 \times 10^{12} \, \text{s}^{-1}$.

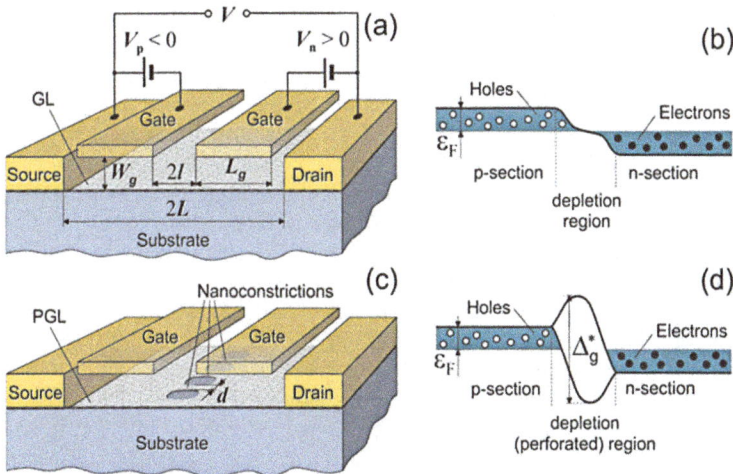

Fig. 7. Schematic views of the split-gate GL-FET (a) and PGL-FET (c) structures and their band digrams (b) and (d), respectively (at unbiased p–n junction).[14]

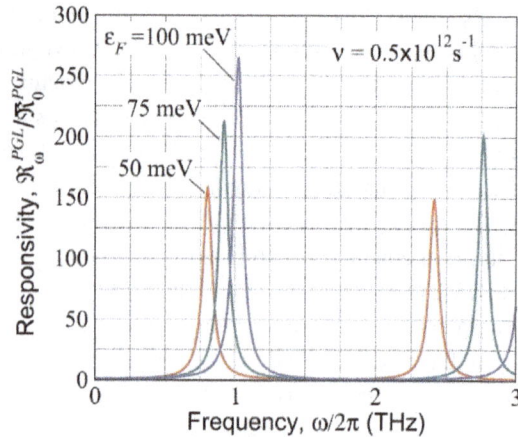

Fig. 8. The PGL-FET responsivity versus signal frequency at different values of Fermi energy shown in figure.[14]

Acknowledgments

The work at RIEC and UoA was supported by the Japan Society for Promotion of Science (KAKENHI Grants No. 23000008 and No. 16H06361), work of RIEC with UB was supported by the RIEC Nation-Wide Cooperative Research Project, and work at RPI was financially supported by Army Research Laboratory under CRA (Project Manager Dr. Meredith Reed).

References

1. M. Dyakonov, and M. S. Shur, "Shallow water analogy for a ballistic field effect transistors: new mechanism of plasma wave generation by dc current", *Phys. Rev. Lett.* **71** (1993) 2465–2468.
2. M. I. Dyakonov and M. S. Shur, "Plasma wave electronics: novel terahertz devices using two dimensional electron fluid", *IEEE Trans. Electron Dev.* **43** (1996) 1640–1645.
3. T. Otsuji, M. Hanabe, and O. Ogawara, "Terahertz plasma wave resonance of two-dimensional electrons in InGaP/InGaAs/GaAs high-electron-mobility transistors", *Appl. Phys. Lett.* **85** (2004) 2119–2121.
4. S. A. Boubanga Tombet, Y. Tanimoto, A. Satou, T. Suemitsu, Y. Wang, H. Minamide, H. Ito, D. V. Fateev, V. V. Popov, and T. Otsuji, "Current-driven detection of terahertz radiation using a dual-grating-gate plasmonic detector", *Appl. Phys. Lett.* **104** (2014) 262104.
5. Y. Kurita, G. Ducournau, D. Coquillat, A. Satou, K. Kobayashi, S. A. Boubanga-Tombet, Y. M. Meziani, V. V. Popov, W. Knap, T. Suemitsu, and T. Otsuji, "Ultrahigh sensitive sub-terahertz detection by InP-based asymmetric dual-grating-gate high-electron-mobility transistors and their broadband characteristics", *Appl. Phys. Lett.* **104** (2014) 251114.
6. V. Ryzhii, M. Ryzhii, N. Ryabova, V. Mitin, and T. Otsuji, "Terahertz and infrared detectors based on graphene structures", *Infrared Phys. and Technol.* **54** (2011) 302–305.
7. V. Ryzhii, N. Ryabova, M. Ryzhii, N. V. Baryshnikov, V. E. Karasik, V. Mitin, and T. Otsuji, "Terahertz and infrared photodetectors based on multiple graphene layer and nanoribbon structures", *Optoelectron. Rev.* **20** (2012) 15–25.

8. V. Ryzhii, T. Otsuji, M. Ryzhii, and M. S. Shur, "Double graphene-layer plasma resonances terahertz detector", *J. Phys. D: Appl. Phys.* **45** (2012) 302001.
9. T. Watanabe, T. Fukushima, Y. Yabe, S. A. B. Tombet, A. Satou, A. Dubinov, V. Y. Aleshkin, V. Mitin, V. Ryzhii and T. Otsuji, "The gain enhancement effect of surface plasmon polaritons on terahertz stimulated emission in optically pumped monolayer graphene", *New J. Phys.* **15** (2013) 075003.
10. V. Ryzhii, A.A. Dubinov, T. Otsuji, V. Ya. Aleshkin, M. Ryzhii, and M. Shur, "Double-graphene-layer terahertz laser: concept, characteristics, and comparison", *Opt. Exp.* **21** (2013) 31569–31579.
11. V. Ryzhii, A. Satou, T. Otsuji, M. Ryzhii, V. Mitin and M. S. Shur, "Dynamic effects in double graphene-layer structures with inter-layer resonant-tunneling negative conductivity", *J. Phys. D: Appl. Phys.* **46** (2013) 315107.
12. V. Ryzhii, T. Otsuji, V. Ya. Aleshkin, A. A. Dubinov, M. Ryzhii, V. Mitin, and M. S. Shur, "Voltage-tunable terahertz and infrared photodetectors based on double-graphene-layer structures", *Appl. Phys. Lett.* **104** (2014) 163505.
13. V. Ryzhii, T. Otsuji, M. Ryzhii, V. Mitin, and M. S. Shur, "Resonant plasmonic terahertz detection in vertical graphene-base hot-electron transistors", *J. Appl. Phys.* **118** (2015) 204501.
14. V. Ryzhii, M. Ryzhii, M. S. Shur, V. Mitin, A. Satou, and T. Otsuji, "Resonant plasmonic terahertz detection in graphene split-gate field-effect transistor with lateral p-n junctions", *J. Phys. D: Appl. Phys.* **49** (2016) 315103.

Sub-Micron Gate Length Field Effect Transistors as
Broad Band Detectors of Terahertz Radiation

J. A. Delgado Notario, E. Javadi[*], J. Calvo-Gallego, E. Diez, J. E. Velázquez,
and Y. M. Meziani[1]

*NanoLab, Salamanca University, Pza. de la Merced, Edificio Trilingue,
Salamanca, 37008, Spain*
[1]*meziani@usal.es*

K. Fobelets

*Department of Electrical and Electronic Engineering, Imperial College,
London, Imperial College London, South Kensington Campus, London SW7 2AZ, UK*

We report on room temperature non-resonant detection of terahertz radiation using strained Silicon MODFETs with nanoscale gate lengths. The devices were excited at room temperature by an electronic source at 150 and 300 GHz. A maximum intensity of the photoresponse signal was observed around the threshold voltage. Results from numerical simulations based on synopsys TCAD are in agreement with experimental ones. The NEP and Responsivity were calculated from the photoreponse signal obtained experimentally. Those values are competitive with the commercial ones. A maximum of photoresponse was obtained (for all devices) when the polarization of the incident terahertz radiations was in parallel with the fingers of the gate pads. For applications, the device was used as a sensor within a terahertz imaging system and its ability for inspection of hidden objects was demonstrated.

Keywords: Si-MODFET; Terahertz detector; Plasma waves.

1. Introduction

The development of novel materials, concepts and device designs for terahertz radiation detection using semiconductors[1] has recently fueled the research of room temperature THz detectors. In early 90's, Dyakonov *et al.*[2,3] theoretically demonstrated the possibility of using sub-micron field effect transistors as detectors of terahertz radiation by means of the oscillations of plasma waves in their channel. Those devices present many advantages: low cost, small size, room temperature operation, and high speed response that make them highly competitive with other technologies. Room temperature detection of sub-terahertz radiation has been demonstrated by using different types of transistors such as commercial GaAs FETs[4] and Si-MOSFET.[5] An array of Si-MOSFETs was

---

*Also with: School of ECE, College of Engineering, University of Tehran, Tehran, Iran.

reported[6] with responsivity ~70 kV/W and NEP (Noise Equivalent Power) ~300 pW/Hz$^{1/2}$. Recently, asymmetric double grating gates devices based on GaInAS/InP HEMTs have shown a record of responsivity and NEP of around 20 kV/W and 0.48 pW/Hz$^{1/2}$, respectively.[7] Inspection feature of terahertz imaging has been demonstrated using those detectors.[5-8] New devices based on topological insulators[9] or graphene[10-11] are also investigated toward higher performance in terms of responsivity and NEP.

In the present work, we investigated room temperature terahertz detection using Strained Silicon Modulation field effect transistor (MODFET) with different gate lengths. Those devices are compatible with the main stream CMOS technology and do show higher mobility in comparison with CMOS ones. Experimental results show a good level of response to terahertz radiation at 150/300 GHz. Simulation results, based on 2D numerical studies using Synopsys TCAD,[12] show a non-resonant response in agreement with measurements. Very competitive values of performance parameters (NEP and responsivity) were obtained. Our device was used as a sensor within a THz imaging system and its capability for inspection of hidden objects was demonstrated.

2. Strained Silicon MODFET and TCAD Modeling

The material system Si/SiGe allows the creation of a thin layer of strained silicon under tetragonal (biaxial tensile) strain. This strain has significant implications for the band structure of the semiconductor.

Tetragonal strain has the effect of lifting the six-fold degeneracy of the conduction band in silicon into a two-fold and four-fold degenerate set. The deformation potential of the strain lowers the energy of the two valleys with their long axis perpendicular to the Si/SiGe interface. The amount of energy lowering is dependent on the degree of strain. It has been theoretically predicted[13] that approximately 0.8 % strain, resulting from a $Si_{0.8}Ge_{0.2}$ alloy, provides sufficient lowering that only the two-fold degenerate valleys are occupied at room temperature for low values of the electric field. Since intervalley carrier scattering may only occur between degenerate minima, electrons in a layer of (tensile) strained silicon would undergo a lower number of intervalley scattering events per unit time – a considerable part of the intervalley scatterings that take place in a bulk material will vanish in strained Si as there are fewer possible final states for a carrier to scatter into – than in a similar layer of bulk silicon. The combination of lower scattering rates (higher values of the momentum relaxation time) and a lower value of the electron conductivity mass as compared to bulk Si makes tensile strained silicon layers excellent candidates to be used in high-mobility FET channels as the transport in those structures is parallel to the Si/SiGe interface.

The epistructure of the MODFETs used in this work was grown by molecular beam epitaxy (MBE) on a thick relaxed SiGe virtual substrate grown by low-energy plasma-enhanced chemical vapor deposition (LEPECVD) over a p-doped conventional Si wafer [Fig. 1(a)]. The final Ge molar concentration in the virtual substrate was $x_{Ge} = 0.45$. The device had a 12 nm tensile strained Si channel, sandwiched between two heavily doped SiGe electron supply layers to generate a high carrier density in the Strained-Si quantum

(a)

3,5 nm Si
6 nm SiGe 30%
5 nm n-SiGe 30% : Sb 01,5E19
4 nm SiGe 30% Spacer
12 nm Si channel
3 nm SiGe 30% Spacer
5 nm n-SiGe 30% : Sb 2E18
100 nm SiGe 30%
200 nm Helax VS SiGe 33%
Substrate: p- > 1000 Ohmcm

(b)

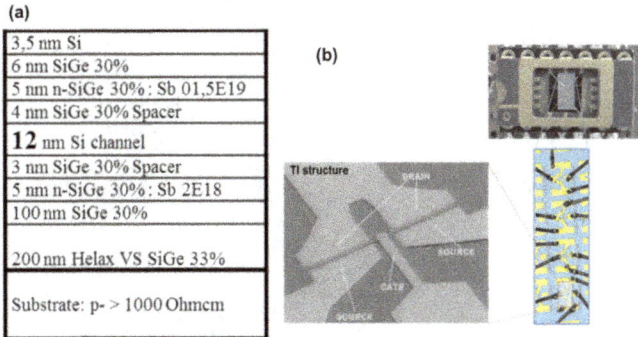

Fig. 1. (a) Epi-structure of devices (b) T-shape design and devices on DIL14 with a microscope image of the devices.

well;[12] to ensure high channel mobility spacers were used to reduce remote impurity scattering by dopants in the supply layers. The ohmic contacts were not self-aligned, the gate electrode was not symmetrically placed between source and drain. All devices had a T shape design [Fig. 1(b)] and were mounted on the same dual inline package (DIL14). Table 1 shows the geometrical parameters of three transistors; typical transfer character-istics are presented in the inset of Figs. 3 and 4.

Owing to the nature of the carrier transport in submicron length gate devices, we used a hydrodynamic (2DHD) model for the majority carriers (electrons) self-consistently coupled to a two-dimensional solution of the Poisson equation.[14] Transport parameters for both holes and electrons were obtained by fitting uniform-field Monte Carlo results obtained for unstrained SiGe and for Si under tensile strain. In the 2DHD simulations impurity de-ionization, Fermi-Dirac statistics and mobility degradation due to both longitudinal and transverse electric field were taken into account. The source and drain regions were simulated as non-self-aligned implanted contacts.

Table 1. Geometrical parameters of the SiMODFETs.

	L_{SD} (μm)	L_g (nm)	W_{SD} (μm)
Device 1	2	100	80
Device 2	2	150	80
Device 3	2	500	80

The study of the THz photovoltaic response of the transistor is implemented by biasing source and gate and floating the drain contact while, as in measurements, a sub-THz sinusoidal signal was superimposed to the gate voltage as described in Refs. 2 and 3. The amplitude of the gate signal was fixed to 5 mV; the induced drain voltage exhibit both the same shape (sinusoidal) and frequency of the gate AC voltage ensuring that no frequency conversion takes place but its amplitude is considerably smaller than the one of the gate's sine wave as in the sub-THz range the device is unable to amplify signals, additionally the mean value of the induced drain voltage were negative in good agreement with theoretical models.[2,3]

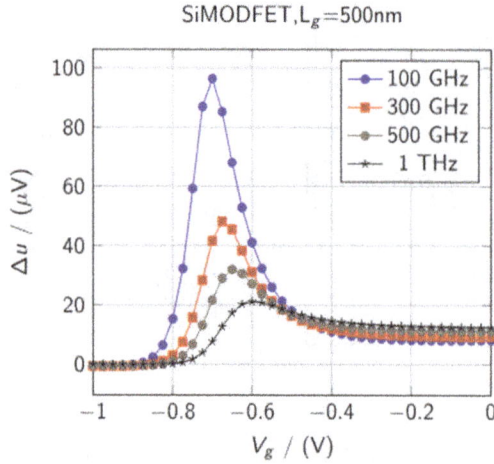

Fig. 2. Simulated photoresponse signal as a function of gate voltage for a device with L_g = 500 nm and under different frequency excitation (100, 300, 500 GHz, and 1 THz).

The charge boundary condition for the floating electrode was specified as:

$$\oint \vec{D} \, \overrightarrow{dS} = Q, \tag{1}$$

where $\vec{D}$ is the displacement vector, Q is the total charge and the integral is evaluated over the drain contact surface. A time-domain simulation was subsequently carried out to obtain the photovoltaic response (Fig. 2).

3. Results and Discussions

The devices were excited at room temperature by a dual-frequency electronic source based on frequency multipliers. The emission frequencies were 150 and 300 GHz with output power levels of 4 mW and 6 mW, respectively. The incoming radiation intensity was modulated by a mechanical chopper between 0.233 to 1.29 kHz and coupled to the device via the metallization pads and bounding wires. The induced photoresponse (Δu) signal was measured by using the lock-in technique. Typical obtained photoresponse signals vs. gate voltage are shown in Fig. 3 for device 1 with L_g = 100 nm under excitation of 150 (blue square symbol) and 300 GHz (blue dotted symbol). Figure 4 shows the case of device 3 with L_g = 500 nm under excitation at 150 GHz (blue square symbol) at 300 GHz (blue dotted symbol). In both cases, it's clearly seen the high value of the signal to noise ratio was obtained and a maximum intensity was observed around the threshold voltage. This behavior has been reported earlier and explained as non-resonant (broadband) detection due to low quality factor ($Q = \omega\tau < 1$)[15] i.e. low mobility in the device's channel. In the present case, the device shows higher channel mobility (~1300 cm^2/V·s) as compared to the conventional Si-MOSFET (~200 cm^2/V·s) the quality factor was estimated around 0.13 at f = 150 GHz and ~0.27 at f = 300 GHz and the resonance condition is not observed.

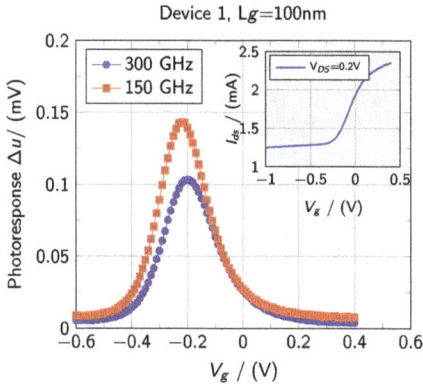

Fig. 3. Photoresponse vs. gate voltage under excitation of 150 GHz (red square) and 300 GHz (blue dotted) for Device 1 with L_g = 100 nm. Inset shows the corresponding transfer characteristics.

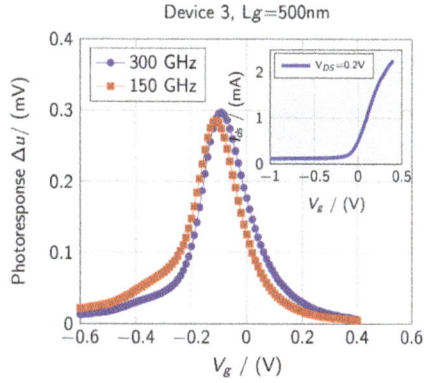

Fig. 4. Photoresponse vs. gate voltage under of 150 GHz (red square) and 300 GHz (blue dotted) for Device 3 with L_g = 500 nm. Inset shows the corresponding transfer characteristics.

Responsivity as well as the Noise Equivalent Power (NEP) are the key parameters to determine the performance of this detectors. The NEP is given by: N_{th}/R_V, where N_{th} is the thermal noise of the transistor in $V/Hz^{0.5}$ and R_V is the responsivity in V/W. Since detection was studied at zero drain bias the thermal noise $N_{th} = (4kTR_{ds})^{0.5}$ was the only source of noise of the transistor. Here R_{ds} is the drain-to-source resistance, which was extracted from the transfer characteristics measured at low drain bias corresponding to the linear regime (inset of Figs. 3 and 4). The responsivity is given as: $R_V = \Delta u \cdot S_t /(P_t \cdot S_a)$, where Δu is the measured photoresponse, S_t is the radiation beam spot area, S_a the active area, and P_t the total incident power surrounding the detector. The radiation beam power at the detector position was estimated to be around 0.5/1 mW at 150/300 GHz, respectively. The spot area S_t is given by πr^2 where r is the radius of the beam spot ($\approx$ 1.5 mm @ 300 GHz and 3.3 mm @ 150 GHz). Since the area of both transistors with contact pads [Fig. 1(b)] was much smaller than the diffraction limit area $S_\lambda = \lambda^2/4$, the active area was taken equal to S_λ. Figure 4 shows the NEP = N_{th}/R_V for device 1 under excitation of 150 (blue dotted) and 300 GHz (red square). As seen, the device shows a minimum NEP around the threshold voltage. Therefore, the NEP$_{min}$ and maximum responsivity (R_{v-max}) for this device are: at 150 GHz, NEP$_{min}$ ~ 150 pW/Hz$^{0.5}$, R_{v-max} ~ 10 V/W and at 300 GHz, NEP ~ 500 pW/Hz$^{0.5}$, R_v ~ 3 V/W.

Table 2 summarizes the obtained NEP and R_v for the remaining studied devices at 150 and 300 GHz. Device 2 with L_g = 150 nm shows the best performance at 150 GHz where: R_{v-max} ~ 15V/W @ 150 GHz and NEP$_{min}$ ~ 150 pW/Hz$^{0.5}$. This must be attributed to the large photoresponse signal exhibited by this device and to a better coupling with the incoming terahertz radiation. Those obtained values of NEP and responsivity are comparable to the commercial room temperature terahertz detectors like golay cell, pyroelectric detector, and Schottky diode. However, the Si-MODFET presents the advantage of working at higher values of modulation frequency.

Table 2. Calculated NEPs and R_v for different devices.

	150 GHz		300 GHz	
	NEP (pW/Hz$^{0.5}$)	R_v (V/W)	NEP (pW/Hz$^{0.5}$)	R_v (V/W)
Device 1	158	9.8	520	3
Device 2	125	15	546	3.6
Device 4	400	9.8	430	8.5

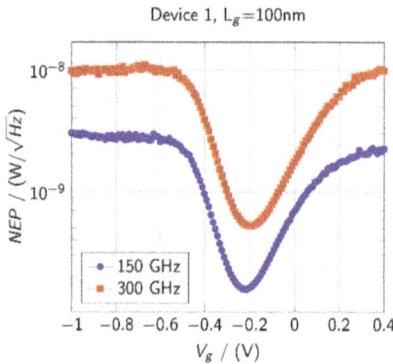

Fig. 5. Noise equivalent power as a function of the gate voltage of Device 1 under different THz excitations: 150 GHz (blue dotted), 300 GHz (red square), $T = 300$ K.

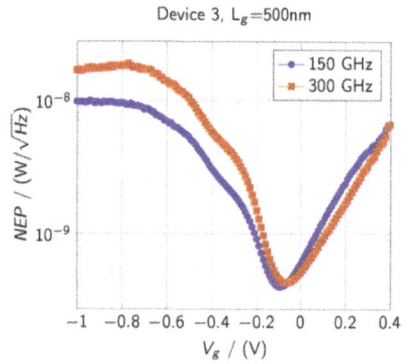

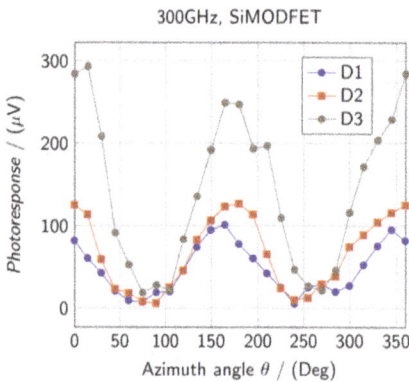

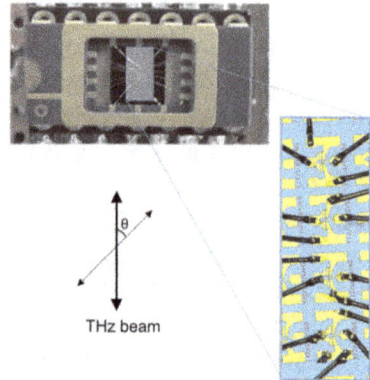

Fig. 6. Noise equivalent power as a function of the gate voltage of Device 3 under different THz excitations: 150 GHz (blue dotted), 300 GHz (red square), $T = 300$ K.

Fig. 7. Photoresponse vs. rotation angle for all devices under excitation of 300 GHz. Inset shows the devices mounted on dual in line package and at 0° where polarisation of THz beam is parallel to the gates pads.

The bonding wires and the metallic pads could play an antenna role to couple the incoming terahertz radiation to the 2D electron channel.[16-18] Sakowicz *et al.*[17] have shown that at low frequencies the radiation is coupled to the transistor mainly by bonding wires, whereas at higher frequencies (> 100 GHz) the metallization of the contact pads plays the role of efficient antennas. Figure 7 shows the photoresponse signal as a function of the polarization of the incoming THz radiation (linearly polarized). In measurements, the device was rotated in the plane perpendicular to the terahertz beam and the photoresponse signal was measured for each angular position of the device. All devices were mounted on the same dual in line package (Fig. 7) and under excitation of 300 GHz. A maximum intensity signal was obtained around the threshold bias for each angle. It is clearly seen for all the devices, that a maximum of the signal results when the incoming radiation is parallel to the gate fingers pads. We can conclude that to maximize the detection of THz signals these must be polarized in a direction parallel to the gate pads.

4. Terahertz Imaging

To demonstrate its usefulness for practical applications, the detector was used as a sensor within a terahertz imaging system. The emitted radiation from the source at 150 and 300 GHz was collimated and focused through different off-axis parabolic mirrors. The image was collected in transmission mode. A visible red LED in combination with an indium tin oxide (ITO) mirror are used for the alignment of the terahertz beam. More information about the terahertz imaging system setup can be found in Ref. 8. The radiation passes through the hidden object and the intensity measured by the device biased around the threshold voltage for maximum intensity of the signal. Figure 8 shows the terahertz images obtained at 150 and 300 GHz as well as the visible one. Better resolution was obtained at 300 GHz which is related to its lower wavelength ($\lambda = 1$ mm). A clear terahertz image of the inspected object was obtained.

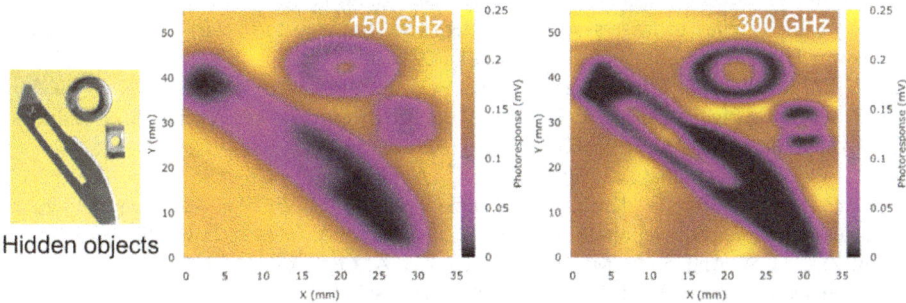

Fig. 8. Visible image (left) and terahertz ones (right) at 150/300 GHz obtained at room temperature with Device 3 with $L_g = 500$ nm.

5. Conclusion

We report on non-resonant detection of terahertz radiation at 150 and 300 GHz at room temperature by using strained-Si MODFETs with different gate lengths. Simulation results, based on 2D numerical studies with Synopsys TCAD, show a non-resonant response in agreement with measurements. Competitive values of the performance parameters (NEP and responsivity) were obtained. We have shown that in this case the gate pads are playing the role of antennae to couple the THz radiation to the 2D electron channel. The practical use of those detectors, a terahertz imaging for inspection of hidden objects was demonstrated.

Acknowledgments

This work was financially supported by the MINECO/FEDER Research Grant #TEC2015-65477-R and FEDER/Junta de Castilla y León Research Grant #SA045U16. Authors thank Dr. W. Knap and D. But from Teralab at Montpellier 2 University for calibration of our detectors. Y. M. Meziani acknowledges the financial support from RIEC nation-wide Collaborative Research Project, Sendai Japan.

References

1. M. Tonouchi, "Cutting-edge terahertz technology", *Nat. Photon.* **1** (2007) 97–105.
2. M. Dyakonov and M. S. Shur, "Shallow water analogy for a ballistic field effect transistor: New mechanism of plasma wave generation by dc current", *Phys. Rev. Lett.* **71** (1993) 2465.
3. M. Dyakonov and M. S. Shur, "Mixing, and frequency multiplication of terahertz radiation by two dimensional electronic fluid", *IEEE Trans. Electron Dev.* **43** (1996) 380.
4. W. Knap, Y. Deng, S. Rumyantsev, J. Q. Lu, M. S. Shur, C. A Saylor, and L. C. Brunel, "Resonant detection of subterahertz radiation by plasma waves in a submicron field-effect transistor", *Appl. Phys. Lett.* **80** (2002) 3433.
5. R. Tauk, F. Teppe, S. Boubanga, D. Coquillat, W. Knap, Y. M. Meziani, C. Gallon, F. Boeuf, T. Skotnicki, C. Fenouillet-Beranger, D. K. Maude, S. Rumyantsev, and M. S. Shur, "Plasma wave detection of terahertz radiation by silicon field effects transistors: Responsivity and noise equivalent power", *Appl. Phys. Lett.* **89** (2006) 253511.
6. A. Lisauskas, U. Pfeiffer, E. Öjefors, P. H. Bolívar, D. Glaab, and H. G. Roskos, "Rational design of high-responsivity detectors of terahertz radiation based on distributed self-mixing in silicon field-effect transistors", *J. Appl. Phys.* **105** (2009) 114511.
7. T. Otsuji, "Trends in the research of modern terahertz detectors: plasmon detectors", *IEEE Trans. Terahertz Sci. Technol.* **5(6)** (2015) 1110–1120.
8. Y. M. Meziani, E. García-García, J. E. Velázquez-Pérez, D. Coquillat, N. Dyakonova, W. Knap, I. Grigelionis, and K. Fobelets, "Terahertz imaging using strained-Si MODFETs as sensors", *Solid-State Electron.* **83** (2013) 113–117.
9. L. Viti, D. Coquillat, A. Politano, K. A. Kokh, Z. S. Aliev, M. B. Babanly, O. E. Tereshchenko, W. Knap, E. V. Chulkov, and M. S. Vitiello, "Plasma-wave terahertz detection mediated by topological insulators surface states", *Nano Lett.*, **16** (2016) 80–87.
10. A. Zak, M. A. Andersson, M. Bauer, J. Matukas, A. Lisauskas, H. G. Roskos, and J. Stake, "Antenna-integrated 0.6 THz FET direct detectors based on CVD graphene", *Nano Lett.* **14(10)** (2014) 5834–5838.
11. L. Vicarelli, M. S. Vitiello, D. Coquillat, A. Lombardo, A. C. Ferrari, W. Knap, M. Polini, V. Pellegrini, and A. Tredicucci, "Graphene field-effect transistors as room-temperature terahertz detectors", *Nat. Mater.* **11(10)** (2012) 865–871.
12. Synopsys, Taurus User Guide, Version Z-2007.3 Synopsys Inc., Mountain View, CA (2007).
13. S. Takagi, J. L. Hoyt, J. J. Welser, and J. F. Gibbons, "Comparative study of phononlimited mobility of two dimensional electrons in strained and unstrained Si metal–oxide–semiconductor field effect transistors", *J. Appl. Phys.* **80** (1996) 1567-1577.
14. J. E. Velazquez, K. Fobelets, and V. Gaspari, "Study of current fluctuations in deep-submicron Si/SiGe n-channel MOSFET: impact of relevant technological parameters on the thermal noise performance", *Semicond. Sci. Technol.* **19** (2004) S191–S194.
15. W. Knap, F. Teppe, Y. Meziani, N. Dyakonova, J. Lusakowski, F. Boeuf, T. Skotnicki, D. Maude, S. Rumyantsev, and M. S. Shur, "Plasma wave detection of sub-terahertz and terahertz radiation by silicon field-effect transistors", *Appl. Phys. Lett.* **85** (2004) 675–677.

16. C. Drexler, N. Dyakonova, P. Olbrich, J. Karch, M. Schafberger, K. Karpierz, Y. Mityagin, M. B. Lifshits, F. Teppe, O. Klimenko, Y. M. Meziani, W. Knap, and S. D. Ganichev, "Helicity sensitive terahertz radiation detection by field effect transistors", *J. Appl. Phys.* **111** (2012) 124504.

17. M. Sakowicz, J. Lusakowski, K. Karpierz, M. Grynberg, W. Gwarek, S. Boubanga, D. Coquillat, W. Knap, A. Shchepetov, and S. Bollaert, "A high mobility field-effect transistor as an antenna for sub-THz radiation", *AIP Conf. Proc.* **1199** (2010) 503–504.

18. D. B. Veksler, A. V. Muravjov, V. Yu. Kachorovskii, T. A. Elkhatib, K. N. Salama, X.-C. Zhang, and M. S. Shur, "Imaging of field-effect transistors by focused terahertz radiation", Solid-State Electron. **53(6)** (2009) 571–573.

Improved Performance of Ultrahigh-Sensitive Charge-Sensitive Infrared Phototransistors (CSIP)

Sunmi Kim

Institute of Industrial Science, The University of Tokyo,
4-6-1, Komaba, Meguro, Tokyo, 153-8505, Japan
kimsunmi@iis.u-tokyo.ac.jp

Susumu Komiyama[1] and Shinpei Matsuda[2]

Department of Basic Science, The University of Tokyo,
Komaba 3-8-1, Meguro-ku, Tokyo, 153-8902, Japan
[1]csusukom@mail.ecc.u-tokyo.ac.jp
[2]s.matsuda1985@gmail.com

Mikhail Patrashin

Frontier Research Laboratory, National Institute of Information and Communications Technology,
4-2-1, Nukui-Kitamachi, Koganei, Tokyo, 184-8795, Japan
mikhail@nict.go.jp

Yusuke Kajihara

Institute of Industrial Science, The University of Tokyo,
4-6-1, Komaba, Meguro, Tokyo, 153-8505, Japan
kajihara@iis.u-tokyo.ac.jp

Charge-sensitive infrared phototransistor (CSIP) is a highly sensitive semiconductor terahertz (THz) detector with single-photon sensitivity. Due to the excitation mechanism via intersubband transition in a quantum well, the CSIP requires careful design of the photo-coupler and proper light illumination method to achieve high quantum efficiency. We have improved the quantum efficiency by introducing the radiation from the backside of the CSIP substrate, which leads to efficient surface plasmon excitation in the photo-coupler.

Keywords: Quantum well based charge-sensitive phototransistor; plasmonic antenna; backside illumination

1. Introduction

Sensitive detection of terahertz (THz) radiation has become increasingly important in a variety of fields including material science, biology, and astrophysics.[1] Among many of THz detectors, the charge-sensitive infrared phototransistor (CSIP) is featured by its unique detection scheme, and is distinguished by its ultra-high sensitivity allowing single photon detection.[2,3] Usually the device consists of a double quantum well (QW) structure, in which photo-generated holes in a floating gate (upper QW: UQW) are sensed by the conductance change in a capacitively-coupled source-drain channel (lower QW: LQW), the effect being similar to that of a field effect transistor (refer Fig. 1). The detection scheme is different from the widely used familiar QW infrared photodetector (QWIP), which works as a photo-diode in which photo carriers are driven across a stack of several tens of QWs.[4] In CSIP, positive charges photo-generated in the floating gate are long lived, leading to a large amplification effect and a huge current responsivity ($> 10^6$ A/W). This results in an extremely low noise equivalent power (NEP = 7×10^{-20} W/Hz$^{1/2}$) and a high detectivity (1×10^{16} cm Hz$^{1/2}$/W).[3] The unprecedented sensitivity has led to the first realization of passive terahertz near-field microscopy.[5]

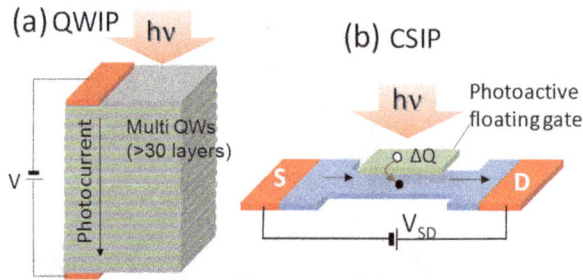

Fig. 1. (a) Schematic view of the conventional quantum well infrared photodetector (QWIP), in which electrons are photoexcited in multiple QWs to yield photocurrent. The photocurrent is driven vertically to the QWs. (b) Charge-sensitive infrared phototransistor (CSIP), in which photo-induced current is driven along the lower QW in the double QW structure. Here the isolated upper QW is charged up by photoexcitation and then the induced charge is detected by the lower QW via capacitive coupling.[2,3]

Among excellent figure of merits of CSIPs, the quantum efficiency (η) is left for challenge: $\eta \sim 7$ % has been reported in a device with metallic cross-hole array antenna for the detection wavelength 14.7 μm,[6] and recently it has been improved up to ~15 % with the optimized cross-hole array antenna for the detection wavelength 16.3 μm. However this value is still less than typical values $\eta > 50$ % reported for QWIP.[7] Improvement of the quantum efficiency is highly desirable for expanding the range of promising application. The difficulty with CSIP is that photons are absorbed via intersubband (ISB) transition of electrons only in a single QW layer located at ca. 100 nm below the crystal surface. The challenge is hence to design an efficient photo-coupler that produces strong electric-field components normal to the QW plane near the crystal surface. Since the photoresponse of CSIP has been improved with the metallic photo-couplers utilizing surface plasmon polariton (SPP) excitation, it should be also noticed that such plasmonic enhancement of CSIP depends on the light incidence method, i.e., frontside versus backside illumination; the former implies the irradiation from air-side on top of CSIP to a GaAs substrate-side, and the latter is for the opposite case as shown in Fig. 2(a). In general, when the incident light from one medium (air) to another (GaAs) through a metallic cross-hole array antenna (Au), the SPPs on metallic photo-coupler are excited on both sides of the metal, namely, SPP_{air} at air/Au boundary and SPP_{GaAs} at Au/GaAs boundary.[8] Then more efficient excitation of SPP_{GaAs} which is important for enhancing the ISB transition of QW is expected for backside illumination. By means of a numerical simulation using a finite-difference time-domain (FDTD) method for the CSIP with cross-hole array antenna, we have found about 3.4 times enhancement on the effective electric field intensity for the backside illumination.[9] Therefore, it is important for further improvement of quantum efficiency of CSIP to study experimentally the effect of backside illumination. In this manuscript, we report experimental results of photoresponse of CSIP under the front and the backside illumination.

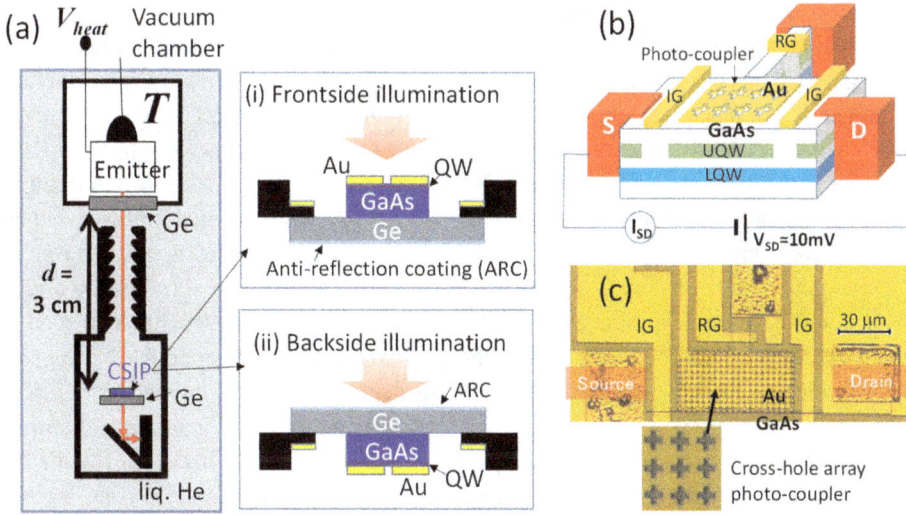

Fig. 2. (a) A diagram of the experimental setup for a measurement of quantum efficiency. The insets show the configurations of (i) frontside and (ii) backside illumination. In order to minimize reflection at GaAs substrate/air interface for backside illumination, an anti-reflection coated Ge plate is attached below CSIP device. (b) Schematic view of CSIP device and (c) a photograph of the fabricated CSIP.

2. Experimental

Samples are fabricated in GaAs/AlGaAs double QW heterostructures grown by molecular-beam epitaxy.[10] CSIP devices with a plasmonic cross-hole array antenna are fabricated via electron beam lithography and wet chemical etching, as schematically shown in Figs. 2(b) and 2(c). Here the pitch of cross-hole array is about 3.5 μm (close to the wavelength in GaAs, i.e., λ/n_{GaAs} for the target wavelength 11.3 μm and the refractive index of GaAs n_{GaAs}).[6] Ohmic contact for source (S) and drain (D) electrodes are formed by alloying with AuGeNi. Isolation gate (IG) and reset gate (RG) are formed by depositing 120 nm-thick Ti/Au layer. The photoactive area of $35{\times}77$ μm^2 is defined by the negatively biased IG. The detection wavelength is experimentally determined to be 11.3 μm, in agreement to the theoretical value of the ISB transition energy obtained from simulation tool "*nextnano*" for the UQW. All the measurements are made at $T = 4.2$ K. Absolute magnitude of quantum efficiency η is derived by measuring the rate of photon detection against the incident photon flux from a temperature-controlled chip resistor, which serves as a black-body radiation source.[11] The experimental setup is shown in Fig. 2(a), where the incident photon flux Φ is determined by monitoring the temperature of the chip resistor via a thermo-couple.

3. Results and Discussion

In order to investigate the quantum efficiency of CSIP, the rate of photon detection (i.e., the count rate) is experimentally derived from the time traces of the photo-induced change in the source-drain current ΔI_{SD} as shown in Fig. 3, where the time traces are plotted for different emitter temperatures. Two configurations are applied for the mode of illumination, viz., (i) the frontside illumination and (ii) the backside illumination. One finds that ΔI_{SD} in the backside illumination in Fig. 3(b) rises more rapidly with time to be saturated faster than the one in the frontside illumination in Fig. 3(a). The quantum efficiency is defined by the ratio of the photon-count rate α to the incident photon flux Φ, viz., $\eta = \alpha/\Phi$. Here α is experimentally derived from $(\Delta I_{SD}/\Delta t)/\Delta I_e$ with $\Delta I_{SD}/\Delta t$ being the slope of the photo-induced current and $\Delta I_e = 0.4934$ pA being the single-photon-induced unit current increment theoretically determined by the device geometry.[3,11] From the experimental results of Fig. 3, we obtain $\eta \approx 12$ % for the frontside illumination and $\eta \approx 26$ % for the backside illumination. Thus the backside illumination has achieved the highest value of η, which corresponds to the improvement by a factor about 2.1 compared to the frontside illumination in the same CSIP device. The experimental results in our work are consistent with recent experiments reporting that, in quantum dot infrared photodetectors with metal photonic crystal as a photo-coupler, backside illumination yields higher photoresponse by a factor more than 2 compared to the frontside illumination.[8,12]

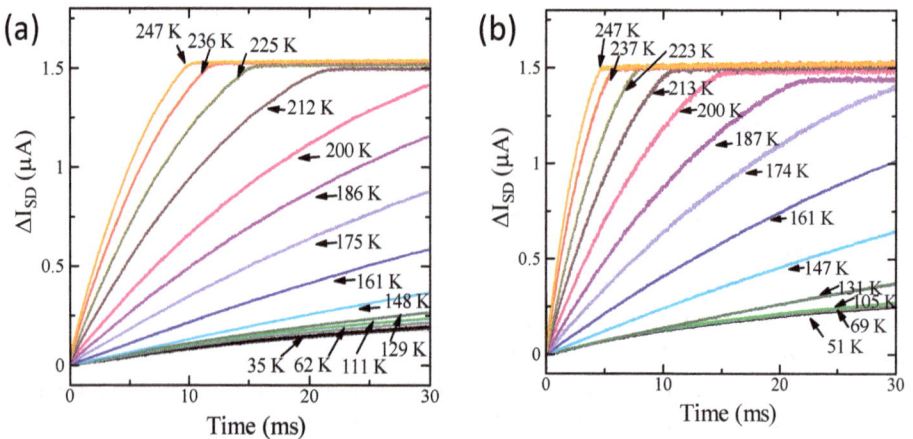

Fig. 3. Time traces of photoresponse under (a) the frontside illumination and (b) the backside illumination with different temperatures of the calibrated blackbody source.

4. Conclusion

We studied efficiency of coupling between incident photons and the electrons in a QW in two different configurations, the frontside illumination and the backside illumination. The quantum efficiency of CSIP has been found to increase from ~12 % to ~26 % in the backside illumination. The enhancement in the backside illumination is interpreted to be a consequence that the SPP is more efficiently excited on the side of GaAs layer when illuminated from the backside.

Acknowledgments

This work is supported by grants for Collaborative Research Based on Industrial Demand from the Japan Science and Technology Agency (JST) and the grant-in-aid for Scientific Research (KAKENHI) C from the Japan Society for the Promotion of Science (JSPS).

References

1. M. Tonouchi, "Cutting-edge terahertz technology", *Nat. Photon.* **1** (2007) 97–105.
2. Z. An, J.-C. Chen, T. Ueda, S. Komiyama, and K. Hirakawa, "Infrared phototransistor using capacitively coupled two-dimentional electron gas layers", *Appl. Phys. Lett.* **86**, 172106, (2005); T. Ueda and S. Komiyama, "Novel ultra-sensitive detectors in the 10–50 μm wavelength range", *Sensors* **10** (2010) 8411–8423.
3. S. Komiyama, "Single-photon detectors in the terahertz range", *IEEE J. Sel. Top. Quantum Electron.* **17** (2011) 54–66.
4. B. F. Levine, "Quantum-well infrared photodetectors", *J. Appl. Phys.* **74** (1993) R1–R81.
5. Y. Kajihara, K. Kosaka, and S. Komiyama, "Thermally excited near-field radiation and far-field interference", *Opt. Exp.* **19(8)** (2011) 7695–7704.
6. P. Nickels, S. Matsuda, T. Ueda, Z. An, and S. Komiyama, "Metal hole arrays as resonant photo-coupler for charge sensitive infrared phototransistors", *IEEE J. Quantum Electron.* **46**, (2010) 384–390.
7. L. Lundqvist, J. Y. Andersson, Z. F. Paska, J. Borglind, and D. Haga, "Efficiency of grating coupled AlGaAs/GaAs quantum well infrared detectors", *Appl. Phys. Lett.* **63** (1993) 3361.
8. S. C. Lee, S. Krishna, and S. R. J. Brueck, "Light direction-dependent plasmonic enhancement in quantum dot infrared photodetectors", *App. Phys. Lett.* **97** (2010) 021112.
9. S. Matsuda, Master thesis, Dept. of Bas. Sci., Univ. Of Tokyo (2012).
10. T. Ueda, Y. Soh, N. Nagai, S. Komiyama, and H. Kubota, "Charge-sensitive infrared phototransistors developed in the wavelength range of 10–50 μm", *Jpn. J. Appl. Phys.* **50** (2011) 020208.
11. T. Ueda, Z. An, K. Hirakawa, and S. Komiyama, "Charge-sensitive infrared phototransistors: Characterization by an all-cryogenic spectrometer", *J. Appl. Phys.* **103** (2008) 093109.
12. G. Gu, J. Vaillancourt, P. Vashinajindakaw, and X. Lu, "Backside-configured surface plasmonic structure with over 40 times photocurrent enhancement", *Semicond. Sci. Technol.* **28(10)** (2013) 105005.

Terahertz Quantum-Cascade Laser
Based on the Resonant-Phonon Depopulation Scheme

Rustam A. Khabibullin[*], Nikolay V. Shchavruk, Aleksandr Yu. Pavlov, Alexey N. Klochkov, Dmitry S. Ponomarev, Igor A. Glinskiy, and Petr P. Maltsev

*Institute of Ultra High Frequency Semiconductor Electronics of RAS,
Moscow, 117105, Russia
[*]khabibullin @isvch.ru*

Alexey E. Zhukov, George E. Cirlin, and Zhores I. Alferov

*St Petersburg National Research Academic University of RAS,
St Petersburg, 194021, Russia*

We have designed GaAs/$Al_{0.15}Ga_{0.85}As$ multilayer heterostructure (MH) with diagonal transitions and optimized oscillator strength – 0.425. We investigate the dependence of the MH energy band profile on the electric field and thermal properties of terahertz quantum cascade lasers (THz QCL) under different operation conditions. Furthermore, we develop a technique for the fabrication of THz QCL with double metal waveguide via low-temperature In-Au wafer bonding followed by substrate removal. Inductively coupled plasma reactive ion etching in BCl_3/Ar at 15:15 sccm has been used to obtain ridge structure of various widths with vertical sidewalls.

Keywords: terahertz quantum-cascade laser; THz source; resonant-phonon depopulation scheme; diagonal transitions; thermal modelling; energy band calculation.

1. Introduction

Among various techniques for THz generation,[1-3] THz quantum cascade lasers (THz QCL) are compact, coherent, continuous wave (cw) solid-state source with electrical pump, thus, they are considered to be the most promising THz source.

Modern THz QCL based on AlGaAs/GaAs heterostructures operate in the range 1.2-5.0 THz (without the use of strong magnetic fields) in a cw with output powers in excess of 200 mW[4] and in the pulsed mode with a peak powers in excess of 1 W.[5] The spectral width of a THz QCL with distributed-feedback is equal to tens of kHz, it allows the use of THz QCL as a local oscillator for heterodyne detection.[6] In Refs. 7 and 8 frequency comb based on THz QCL with a spectral bandwidth of more than 1 THz was demonstrated, which enables the development of THz spectrometers with a large signal to noise ratio. One of the key desired THz QCL characteristic is room-temperature operation, but it has not been achieved yet.

In the present article we have proposed the design of GaAs/Al$_{0.15}$Ga$_{0.85}$As multilayer heterostructure (MH) with optimized oscillator strength and investigated thermal properties and energy band profile of THz QCL under different operation conditions.

2. Results and Discussion

On the basis of the calculation of the oscillator strength for the transitions between quantum-well levels two MH based on the three-quantum well (3QW) resonant-phonon depopulation (RP) scheme were grown by molecular beam epitaxy. The layer sequences, starting from the injector barrier, are 43/75.6/**24.6**/69.3/**41**/136 (MH-1); **43**/89/**24.6**/81.5/ **41**/160 (MH-2). The results of X-ray diffraction and photoluminescence spectroscopy characterization of the grown MH are presented in Ref. 9.

3QW MH based on RP scheme with four electron states (see Fig. 1) is investigated. Levels E2 and E3 in the double quantum well are active radiative states. Difference E3 – E2 corresponds to the energy and frequency of the emitted THz electromagnetic wave. Levels E1 and E4 in the wide quantum well correspond to the injector and extractor states respectively. The THz QCL generation is possible provided that bias induces following level alignments: injector E1 is aligned with higher laser state E3 and lower laser state E2 is aligned with extractor state E4' in the next period. The heterostructure is designed in such a way that difference E4 – E1 is equal to optical phonon energy to assure resonant electron extraction.

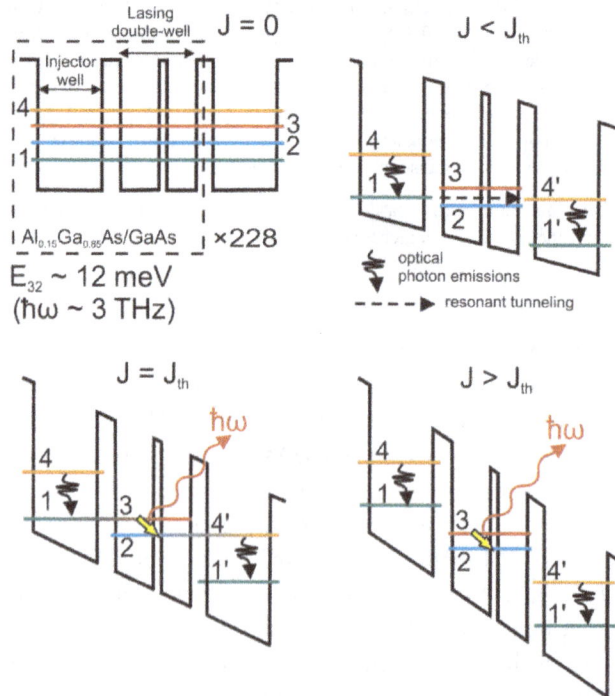

Fig. 1. Conduction-band diagrams of the 3QW MH based on RP scheme at different biases. A single quantum cascade stage is marked by a box. The radiative transition is from 3 to 2.

As seen in Fig. 2 the first level alignment of the lower injector state (E1) with the upper injector state in the next stage (E4') occurs under 8.5 kV/cm (calculated within a Schrodinger-Poisson approach). In that case the electrons tunnel through "lasing double-well" without emission of photons. The second level alignment occurs under 12.5 kV/cm and indicates alignment of the lower injector state (E1) with the upper laser state (E3). At higher bias THz QCL band structure is totally misaligned.

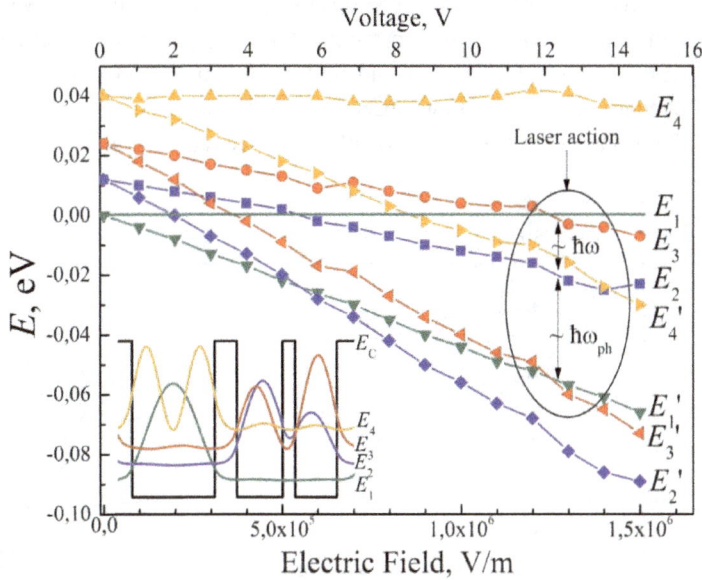

Fig. 2. Dependence of the energy level positions in the current Ei and following Ei' THz QCL periods on electric field for structure MH-2 (level E1 is reference); inset – conduction band scheme, electron wave functions and energy levels alignment at zero electric field.

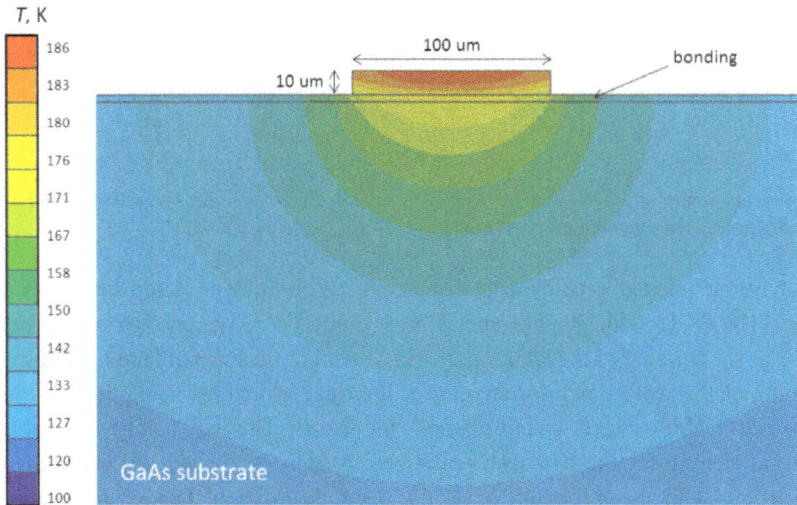

Fig. 3. Two-dimensional heat flow model of cw-THz QCL calculated with finite-element solver. The lower boundary is set to 100 K.

To estimate the thermal properties of THz QCL under the different operation conditions (power and cooling modes) modeling of heat flow processes in these devices was carried out. The results for cw-THz QCL are shown in Fig. 3. Because of the high value of thermal conductivity of the MH and In−Au bonding layer, thermal resistance of the device is dominated by the temperature drop inside the active region, and spreading resistance in the substrate.

The thermal dynamics of a THz QCL operating in pulse mode is investigated. In Fig. 4 maximum values of active region temperature T_{AR} under the different operation conditions (pulse repetition frequencies and duty cycles) are shown. Temperature–time profile for a standard ridge waveguide (inset Fig. 4) showing that T_{max} is stabilized for a time of about 2 ms and the further operation of the device remains unchanged.

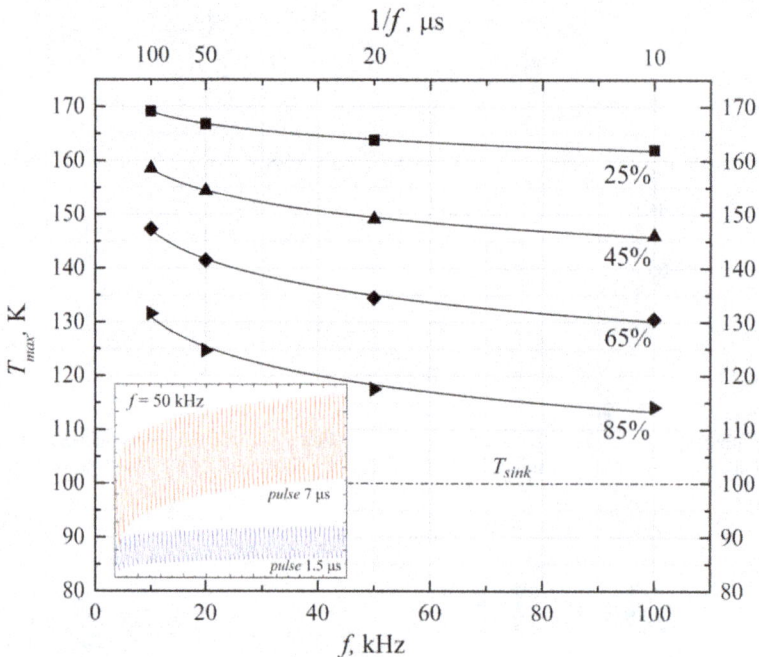

Fig. 4. Maximum values of T_{AR} for different pulse repetition frequencies in the range 10-100 kHz and duty cycle 25-85%; inset – temperature-time profile for 1.5 and 7 μs pulse widths at 50 kHz.

We have developed post-growth processing of multilayer heterostructures GaAs/AlGaAs THz QCL with double metal waveguide.[10,11] The post-growth processing includes In−Au bonding of the QCL on n+-GaAs wafer, mechanical lapping and selective wet etching of the wafer, and dry etching of the ridge structures using 50-μm-wide and 100-μm-wide Ti/Au contacts as self-aligned etch masks. The inductively coupled plasma reactive ion etching regime has been optimized in BCl₃/Ar for vertical sidewalls of the ridge structure with the minimum of Ti/Au mask destruction. Device processing then continued according to the standard recipe.

THz QCL bar is soldered onto Cu blocks, which serves as a heat sink and bottom electrical contact (Fig. 5). Top QCL contact is made by wire bonding to Ti/Au pads. A scanning electron micrograph of a typical wire bonding 50-µm-wide ridge structure is shown in Fig. 6. It has been found that n+-GaAs receptor must be thinned to 120-140 µm to obtain a good quality of laser facet.

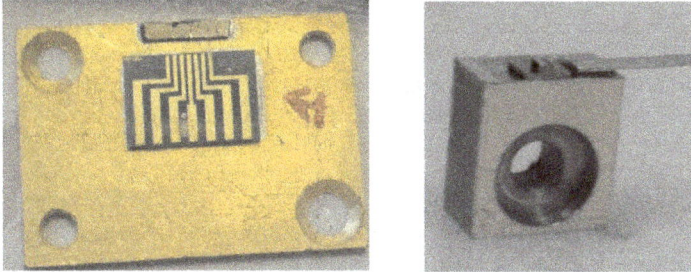

Fig. 5. THz QCL bar is mounted onto Cu blocks.

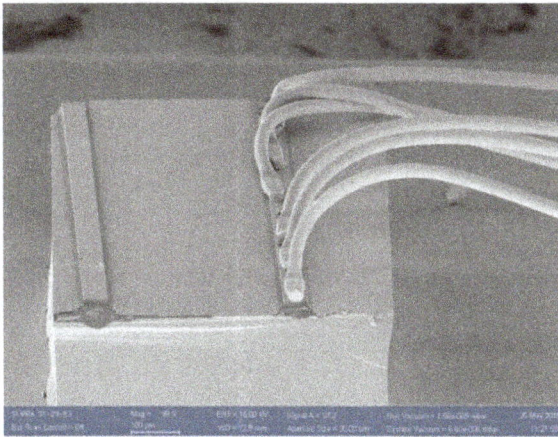

Fig. 6. Scanning electron micrograph of a processed 50-µm-wide dry-etched THz QCL with contact pads.

3. Conclusion

We have proposed the GaAs/Al$_{0.15}$Ga$_{0.85}$As multilayer heterostructure with optimized oscillator strength – 0.425. Energy band calculations and thermal modelling of THz QCL under different operation conditions were carried out. THz QCL with 50 and 100-µm-wide ridge structure were fabricated. Measurements of the given THz QCL are going to be published soon.

Acknowledgments

The work was supported by the Russian President's grant № MK-6081.2016.8.

References

1. R. R. Galiev, A. E. Yachmenev, A. S. Bugaev, G. B. Galiev, Yu. V. Fedorov, E. A. Klimov, R. A. Khabibullin, D. S. Ponomarev, and P. P. Maltsev, "Promising materials for an electronic component base used to create terahertz frequency range (0.5–5.0 THz) generators and detectors", *Bul. Rus. Acad. Sci:. Phys.* **80(4)** (2016) 476–478.

2. D. V. Lavrukhin, A. E. Yachmenev, R. R. Galiev, A. S. Bugaev, Y. V. Fedorov, R. A. Khabibullin, D. S. Ponomarev, and P. P. Maltsev, "Investigation and fabrication of the semiconductor devices based on metamorphic InAlAs/InGaAs/InAlAs nanoheterostructures for THz applications", *Int. J. High Speed Electron. Syst.* **24(1&2)** (2015) 1520001.

3. D. V. Lavrukhin, A. E. Yachmenev, R. R. Galiev, R. A. Khabibullin, D. S. Ponomarev, Yu. V. Fedorov, and P. P. Maltsev, "MHEMT with a power-gain cut-off frequency of f_{max} = 0.63 THz on the basis of a $In_{0.42}Al_{0.58}As/In_{0.42}Ga_{0.58}As/In_{0.42}Al_{0.58}As/GaAs$ nanoheterostructure", *Semiconductors* **48(1)** (2014) 69–72.

4. X. Wang, C. Shen, T. Jiang, Z. Zhan, Q. Deng, W. Li, W. Wu, N. Yang, W. Chu, and S. Duan, "High-power terahertz quantum cascade lasers with ~0.23 W in continuous wave mode", *AIP Adv.* **6** (2016) 075210.

5. L. H. Li, L. Chen, J. X. Zhu, J. Freeeman, P. Dean, A. Valavanis, A. G. Davies, and E. H. Linfield, "Terahertz quantum cascade lasers with >1 Watt output powers", *Electron. Lett.* **50(4)** (2014) 309–311.

6. J. R. Gao, J. N. Hovenier, Z. Q. Yang, J. J. A. Baselmans, A. Baryshev, M. Hajenius, T. M. Klapwijk, A. J. L. Adam, T. O. Klaassen, B. S. Williams, S. Kumar, Q. Hu, and J. L. Reno, "Terahertz heterodyne receiver based on a quantum cascade laser and a superconducting bolometer", *Appl. Phys. Lett.* **86** (2005) 244104.

7. D. Burghoff, T. Y. Kao, N. Han, C. W. I. Chan, X. Cai, Y. Yang, D. J. Hayton, J.-R. Gao, J. L. Reno, and Q. Hu, "Terahertz laser frequency combs", *Nat. Photonics* **8** (2014) 462–467.

8. H. Li, P. Laffaille, D. Gacemi, M. Apfel, C. Sirtori, J. Leonardon, G. Santarelli, M. Rösch, G. Scalari, M. Beck, J. Faist, W. Hänsel, R. Holzwarth, and S. Barbieri, "Dynamics of ultra-broadband terahertz quantum cascade lasers for comb operation", *Opt. Exp.* **23(26)** (2015) 33270–33294.

9. A. E. Zhukov, G. E. Cirlin, R. R. Reznik, Yu. B. Samsonenko, A. I. Khrebtov, M. A. Kaliteevski, K. A. Ivanov, N. V. Kryzhanovskaya, M. V. Maximov, and Zh. I. Alferov, "Multilayer heterostructures for quantum-cascade lasers operating in the terahertz frequency range", *Semiconductors* **50(5)** (2016) 662–666.

10. R. A. Khabibullin, N. V. Shchvruk, A. Yu. Pavlov, D. S. Ponomarev, K. N. Tomosh, R. R. Galiev, P. P. Maltsev, A. E. Zhukov, G. E. Cirlin, F. I. Zubov, and Zh. I. Alferov, "Fabrication of a terahertz quantum-cascade laser with a double metal waveguide based on multilayer GaAs/AlGaAs heterostructures", *Semiconductors* **50(10)** (2016) 1377–1382.

11. R. A. Khabibullin, N. V. Shchavruk, A. N. Klochkov, I. A. Glinskiy, N. V. Zenchenko, D. S. Ponomarev, P. P. Maltsev, A. A. Zaycev, F. I. Zubor, A. E. Zhukov, G. E. Cirlin, Zh. I. Alferov, *Semiconductors* **51(4)** (2017) 540–546.

Intensive Terahertz Radiation from
In$_x$Ga$_{1-x}$As due to Photo-Dember Effect

Dmitry S. Ponomarev*, Rustam A. Khabibullin, Aleksandr E. Yachmenev,
and Petr P. Maltsev†

*Institute of Ultra High Frequency Semiconductor Electronics of RAS,
Moscow, 117105, Russia
*ponomarev_dmitr@mail.ru
†iuhfseras2010@yandex.ru*

Igor E. Ilyakov, Boris V. Shiskin, and Rinat A. Akhmedzhanov

*Institute of Applied Physics of RAS,
Nizhny Novgorod, 603950, Russia*

We have proposed and investigated In$_y$Ga$_{1-y}$As photoconductor grown by molecular-beam epitaxy on low-temperature step-graded metamorphic buffer. It exhibits superior bandwidth up to 6 THz and provides optical-to-terahertz conversion efficiency up to ~ 10^{-5} for rather low optical fluence ~ 40 μJ/cm^2. The intensity of THz generation for the given structure is two orders higher than for low-temperature grown GaAs due substantial contribution of photo-Dember effect.

Keywords: InGaAs; photo-Dember effect; THz radiation; molecular-beam epitaxy; optical-to-terahertz conversion efficiency; metamorphic buffer; time-domain spectroscopy.

1. Introduction

Terahertz (THz) technology has become increasingly popular due to its unique applications in security screening, space exploration, biological sensing and medical imaging.[1,2] Among various techniques for THz generation, photoconductive devices have demonstrated very promising performance for generating both pulsed and continuous-wave THz radiation. THz time-domain spectroscopy (THz-TDS) is widely used in materials science, biology and medicine for investigation of molecules, DNK, RNK, cancer tumors, proteins, etc. THz sources that are used in THz-TDS can be divided into two groups. The first one is attributed to nonlinear conversion of femtosecond laser radiation, and the second is connected with ultrafast dynamics of photoexcited carriers in semiconductors – the so-called photoconductive emitters that use SI-GaAs,[3] low-temperature grown GaAs (LT GaAs)[4,5] and also In$_x$Ga$_{1-x}$As as a photoconductive material.

In$_x$Ga$_{1-x}$As is an attractive candidate for emission of THz radiation at optical pump 1.0–1.6 μm[6,7] where SI or LT GaAs do not work. An ultrafast optical pulse with photon energy above the semiconductor band-gap strikes the semiconductor and creates electron-hole pairs at subsurface of the semiconductor. These pairs are rapidly accelerated by the built-in electric field resulting in a radiative dipole parallel to surface normal.[8,9] The most common used photoconductive THz emitter is photoconductive antenna (PCA) where photoexcited carries are accelerated by applied external bias.[10,11] Large external biases are essential to increase the power spectrum of PCA thus the photoconductive material has to be high resistive. Initially In$_x$Ga$_{1-x}$As has low resistivity and suffers from high dark currents which are detrimental for biased photoconductive devices. There were many attempts to improve it, for instance different authors proposed ErAs islands incorporated into InAlAs/InGaAs quantum wells,[9,10] Be-doped LT InGaAs,[11] ion-implantation in InGaAs[12,13] and etc.

THz generation may occur without an external electrical bias as a result of the boundary conditions on the carrier transport within a semiconductor as a result of the photo-Dember (PD) effect.[14,15] The mechanism of PD effect arises from the differing diffusion mobilities of holes and electrons within a semiconductor. As a rule electron mobility is higher than hole mobility thus under ultrafast optical excitation the spatial distribution of electron-hole pairs occurs leading to THz radiation. PD effect better occurs in narrow-band semiconductors where the difference between electron and hole mobilities is much stronger, for instance in InAs and InN.[16-18] In thermal equilibrium the ratio of the mobilities of electrons in the Γ-valley and heavy holes for In$_x$Ga$_{1-x}$As (x ≥ 0.4) is about 40 and less than 20 for LT GaAs.[19] Thus In$_x$Ga$_{1-x}$As can be used as promising candidate in PD emitters for broadband THz generation. It is important to note that in case of PD effect there is no need of high resistivity of In$_x$Ga$_{1-x}$As and THz emission can scale with the excitation area by multiplexing (repeating) the number of emitters on a wafer.[20,21] In Ref. 19 it was shown that PD emitters provide even higher bandwidth than photoconductive emitters.

The aim of present work is the development of growth technique and investigation of In$_{0.38}$Ga$_{0.62}$As photoconductive material where THz generation is achieved by two mechanisms: 1) acceleration of the photoexcited carriers by the built-in electric field and 2) by PD effect. Due to there is no lattice-matched wafers to In$_x$Ga$_{1-x}$As (0 < x < 0.4) we have proposed the step-graded metamorphic buffer (MB) which allows stoichiometric epitaxial growth of In$_{0.38}$Ga$_{0.62}$As on GaAs wafer.

2. The Samples and Experimental Setup

The sample of In$_{0.38}$Ga$_{0.62}$As was grown by molecular-beam epitaxy (MBE) on (100) GaAs wafer. Its schematic diagram is shown in Fig. 1. The growth temperature of the photoconductive layer In$_{0.38}$Ga$_{0.62}$As was set up to 490 °C and then reduced to 400 °C when MB was grown. The step-graded 1.0 μm MB with inverse step In$_{0.38}$Al$_{0.62}$As consists of seven In$_x$Al$_{1-x}$As layers with x increasing from 0.1 to 0.46. The idea of the given buffer is in step-by-step adjusting of the crystalline parameters of photoconductive layer and GaAs wafer.[22] The inverse step with low indium content decreases elastic strain in the active (photoconductive) area and improves it's structural property.[23,24] The post-growth annealing for In$_{0.38}$Ga$_{0.62}$As grown without access arsenic pressure is no need to be carried out.

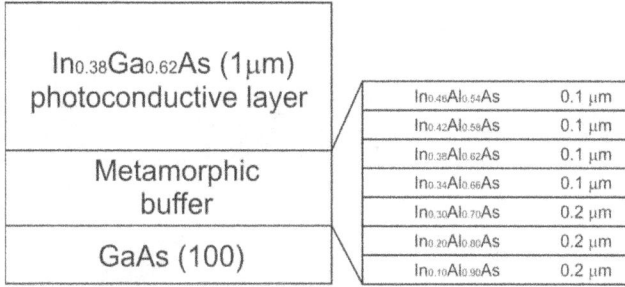

In$_{0.38}$Ga$_{0.62}$As (1 µm) photoconductive layer		
	In$_{0.46}$Al$_{0.54}$As	0.1 µm
	In$_{0.42}$Al$_{0.58}$As	0.1 µm
	In$_{0.38}$Al$_{0.62}$As	0.1 µm
Metamorphic buffer	In$_{0.34}$Al$_{0.66}$As	0.1 µm
	In$_{0.30}$Al$_{0.70}$As	0.2 µm
	In$_{0.20}$Al$_{0.80}$As	0.2 µm
GaAs (100)	In$_{0.10}$Al$_{0.90}$As	0.2 µm

Fig. 1. Schematic diagram of In$_{0.38}$Ga$_{0.62}$As photoconductor grown on step-graded metamorphic buffer.

For comparison we also grew and investigated LT GaAs with heavily doped n-GaAs buffer layer. Previously[5] we have shown that LT GaAs grown by MBE with several thin δ-Si doped layers embedded into the photoconductive layer GaAs exhibits picosecond carrier lifetimes. Authors in Ref. 25 showed that use of additional n-GaAs layer beneath LT GaAs enhances built-in electric field on the interface photoconductive layer/buffer layer and thus increase radiated THz electric field. The growth temperature of LT GaAs was set up to 215 °C. The post-growth annealing was carried out *in situ* at 600 °C during 20 min.

The structural properties were investigated by means of high resolution double crystal X-ray diffraction. Diffraction rocking curves (DRC) were measured on Rigaku Ultima IV.

The samples were investigated and compared by conventional THz-TDS system. They were irradiated by Ti:sapphire mode-locked laser pulses with 795 nm central wavelength and 50 fs pulse duration (pulse energy 800 µJ, aperture 7.0 mm. The generated THz pulses were focused by an off-axis parabolic mirror on a detection crystal [(100) cut 200 µm GaP crystal] and THz electric field was measured by a standard ellipsometric scheme.[26] Carrier lifetimes were measured by using a time-resolved pump-probe technique.

3. Results and Discussion

Figure 2 depicts DRC for In$_{0.38}$Ga$_{0.62}$As measured in θ/2θ-scanning mode (θ is the angle between the reflecting plane and the incident beam, 2θ is the angle between the incident and reflected X-ray beams). It allows determination of the lattice parameters in various directions by the angular positions of peaks on DRC. For symmetric reflections we chose (400) plane.

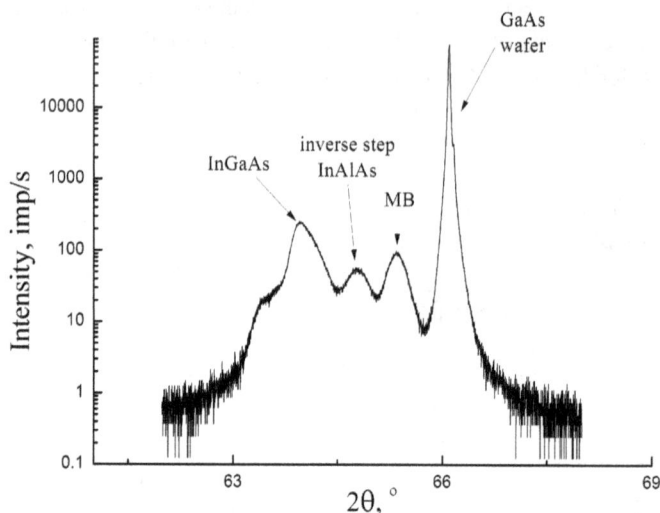

Fig. 2. Diffraction rocking curve of $In_{0.38}Ga_{0.62}As$ grown on step-graded metamorphic buffer.

As seen in Fig. 2 right sharp peak corresponds to (100) GaAs while the left one to photoconductive layer $In_{0.38}Ga_{0.62}As$. Two peaks between them relate to MB and inverse step $In_{0.38}Al_{0.62}As$. We used asymmetric (411) reflections to extract in-plane and out-of-plane lattice parameters for $In_{0.38}Ga_{0.62}As$ (5.8010 Å and 5.8211 Å respectively) and after we calculated the deformation of the given layer that was estimated as $\varepsilon_{res} \sim 0.0015$. This confirms that the layer was grown crystalline with good quality.[27] We shall note that because MB controls threading of defects and dislocations towards active region this provides varying of indium content (x) in $In_xGa_{1-x}As$ layer in wide range up to 100% (*i*-InAs). In other words by means of MB we can alter the bandgap of $In_xGa_{1-x}As$ and adjust it to optical pump 1.0–1.6 μm.

Figure 3 shows optical-pump THz-probe measurements in reflection geometry performed at room temperature with 795 nm pump for $In_{0.38}Ga_{0.62}As$ with MB. Two optical pump fluences of 45 and 580 μJ were used to excite the sample. The penetration depth of 795 nm light into $In_{0.38}Ga_{0.62}As$ is approximately 200 nm, so the 1.0 μm thickness is sufficient to absorb all of the incident pump light. As seen at later time delays, the decay of the signal can be accurately fitted by a single exponential, yielding a carrier lifetime ~10 ps at both optical fluences. The result is not surprising due to $In_{0.38}Ga_{0.62}As$ has good crystalline quality and thus possess much higher electron mobility in comparison to LT InGaAs. Despite this THz-TDS measurements showed that $In_{0.38}Ga_{0.62}As$ with MB provides both high THz bandwidth and intensity of THz generation. Figures 4 and 5 show THz pulse waveforms for $In_{0.38}Ga_{0.62}As$ with MB and LT GaAs in time-domain and Fourier power spectra for both samples respectively. As seen in Fig. 4 the radiation spectrum for $In_{0.38}Ga_{0.62}As$ with MB is two orders higher than for LT GaAs. This is due to much stronger intensity of THz radiation electric field in $In_{0.38}Ga_{0.62}As$ with LT MB (see Fig. 4).

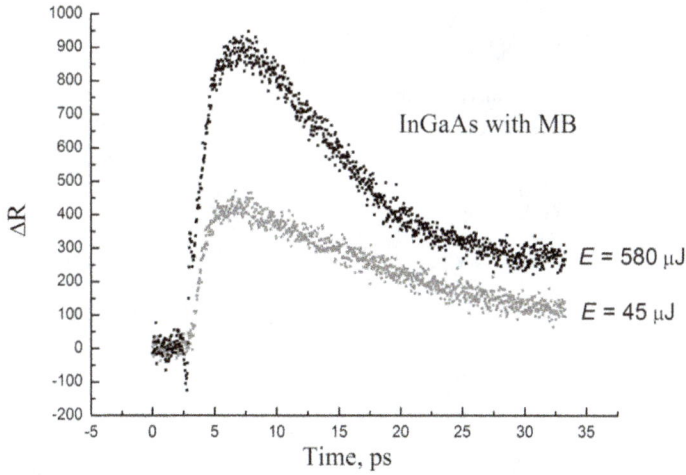

Fig. 3. Optical-pump THz-probe measurements at two pump fluences 45 and 580 μJ with 795 nm pump for In$_{0.38}$Ga$_{0.62}$As with MB.

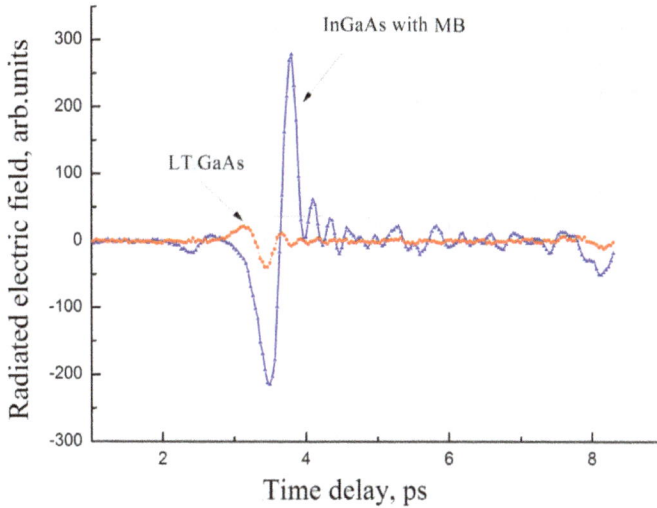

Fig. 4. THz pulse waveforms in time-domain for both samples.

It also demonstrates superior bandwidth up to 6 THz. We assume this is because of PD effect that strongly contributes to THz generation for In$_{0.38}$Ga$_{0.62}$As with MB. Figure 6 indicates THz amplitude dependence on optical pump energy of Ti:S laser for In$_{0.38}$Ga$_{0.62}$As with MB. The amplitude shows linear increase in log-scale with increase of pump energy. As seen at energy pump ~110 μJ the dependence saturates which is connected with finite density of states in In$_{0.38}$Ga$_{0.62}$As[28] and decrease of electron mobility due to enhanced intervalley scattering at high values of pump energy. It should be noted that for rather low optical fluence ~40 μJ/cm^2 the optical-to-terahertz conversion efficiency is ~10^{-5} that is much higher than for LT GaAs.

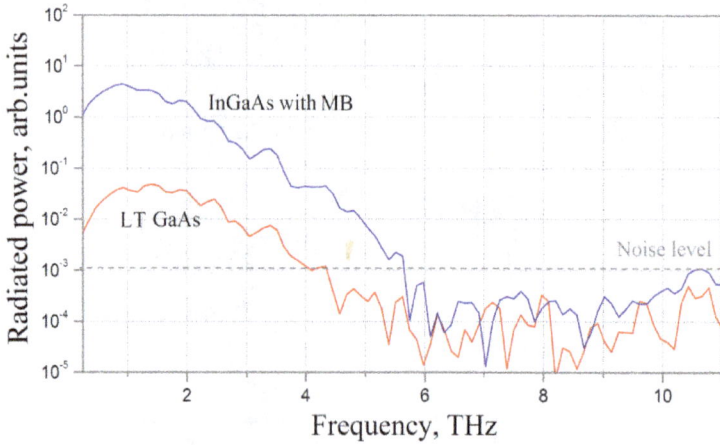

Fig. 5. Fourier power spectra for all the grown samples. Blue line indicates spectrum for InGaAs with MB, red line – for LT GaAs. Grey dotted line is system noise level.

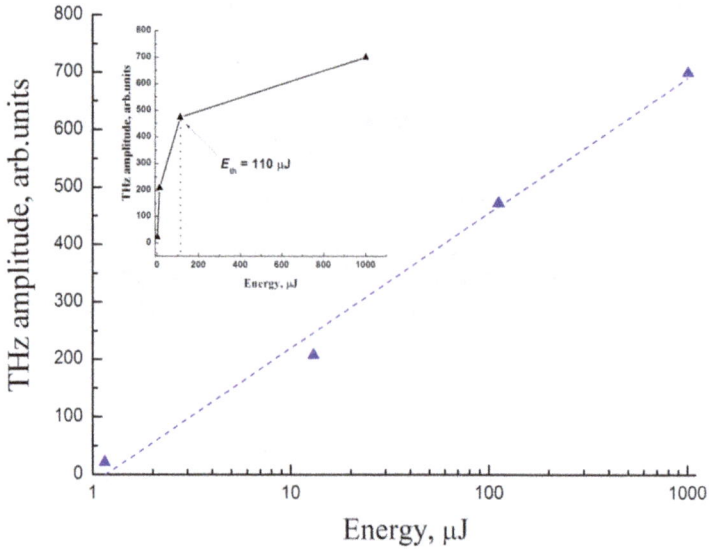

Fig. 6. THz amplitude dependence on optical excitation energy (in log-scale) for InGaAs with MB.

4. Conclusion

In conclusion, we demonstrated that $In_{0.38}Ga_{0.62}As$ photoconductive material exhibits both high THz bandwidth and optical-to-terahertz conversion efficiency and thus can be a promising candidate for use in PCA and photoconductive emitters based on lateral PD effect.[19,28] This topic will form the basis of our future work.

Acknowledgments

The work was supported by the RFBR grants № 16-32-50047, 16-07-00187 A; 16-29-14029 ofi_m and 16-29-03033.

References

1. C. W. Berry, N. Wang, M. R. Hashemi, M. Unlu, and M. Jarrahi, "Significant performance enhancement in photoconductive terahertz optoelectronics by incorporating plasmonic contact electrodes", *Nat. Commun.* **4** (2013) 1622.
2. R. R. Galiev, A. E. Yachmenev, A. S. Bugaev, G. B. Galiev, Yu. V. Fedorov, E. A. Klimov, R. A. Khabibullin, D. S. Ponomarev, and P. P. Maltsev, *Bul. Rus. Academ. Sci.: Phys.* **80(4)** (2016) 476–478.
3. J. T. Kindt and C. A. Schumuttenmaer, "Far-infrared dielectric properties of polar liquids probed by femtosecond terahertz pulse spectroscopy", *J. Phys. Chem.* **100(24)** (1996) 10373–10379.
4. N. T. Yardimci, S.-H. Yang, C. W. Berry, and M. Jarrahi, "High power terahertz generation using large area plasmonic photoconductive emitters", *IEEE Trans. Terahertz Sci. Technol.* **5** (2015) 223–229.
5. D. V. Lavrukhin, A. E. Yachmenev, A. S. Bugaev, G. B. Galiev, E. A. Klimov, R. A. Khabibullin, D. S. Ponomarev, and P. P. Maltsev, "Investigation of the optical properties of GaAs with δ-Si doping grown by molecular-beam epitaxy at low temperatures", *Semiconductors* **49(7)** (2015) 911–914.
6. A. Takazato, M. Kamakura, T. Matsui, J. Kitagawa, and Y. Kadoya, "Terahertz wave emission and detection using photoconductive antennas made on low-temperature-grown InGaAs with 1.56 μm pulse excitation", *Appl. Phys. Lett.*, **91** (2007) 011102.
7. S.-P. Han, H. Ko, N. Kim, H.-C. Ryu, C. W. Lee, Y. A. Leem, D. Lee, M. Y. Jeon, S. K. Noh, H. S. Chun, and K. H. Park, "Optical fiber-coupled InGaAs-based terahertz time-domain spectroscopy system", *Opt. Lett.* **36(16)** (2011) 3094–3096.
8. S. Gupta, J. F. Whitaker, and G. A. Mourou, "Ultrafast carrier dynamics in III-V semiconductors grown by molecular-beam epitaxy at very low substrate temperatures", *IEEE J. Quantum Electron.* **28** (1992) 2464–2472.
9. D. C. Driscoll, M. P. Hanson, A. C. Gossard, and E. R. Brown, "Ultrafast photoresponse at 1.55 μm in InGaAs with embedded semimetallic ErAs nanoparticles", *Appl. Phys. Lett.* **86** (2005) 051908.
10. S. Preu, M. Mittendorf, H. Lu, H. B. Weber, S. Winnerl, and A. C. Gossard, "1550 nm ErAs:In(Al)GaAs large area photoconductive emitters", *Appl. Phys. Lett.* **101** (2012) 101105.
11. D. Vignaud, J. F. Lampin, E. Lefebvre, M. Zaknoune, and F. Mollot, "Electron lifetime of heavily Be-doped In$_{0.53}$Ga$_{0.47}$As as a function of growth temperature and doping density", *Appl. Phys. Lett.* **80(22)** (2002) 4151.
12. N. Chimot, J. Mangeney, P. Mounaix, M. Tondusson, K. Blary, and J. F. Lampin, "Terahertz radiation generated and detected by Br$^+$-irradiated In$_{0.53}$Ga$_{0.47}$As photoconductive antenna excited at 800 nm wavelength", *Appl. Phys. Lett.* **89** (2006) 083519.
13. M. Suzuki and M. Tonouchi, "Fe-implanted InGaAs photoconductive terahertz detectors triggered by 1.56 μm femtosecond optical pulses", *Appl. Phys. Lett.* **86** (2005) 163504.
14. V. Malevich, R. Adomavičius, and A. Krotkus, "THz emission from semiconductor surfaces", *C. R. Physique* **9** (2008) 130–141.
15. M. C. Beard, G. M. Turner, and C. A. Schmuttenmaer, "Subpicosecond carrier dynamics in low-temperature grown GaAs as measured by time-resolved terahertz spectroscopy", *J. Appl. Phys.* **90** (2001) 5915–5923.

16. A. Reklaitis, "Terahertz emission from InAs induced by photo-Dember effect: Hydrodynamic analysis and Monte Carlo simulations", *J. Appl. Phys.* **108** (2010) 053102.

17. K. Liu, J. Z. Xu, T. Yuan, and X.-C. Zhang, "Terahertz radiation from InAs induced by carrier diffusion and drift", *Phys. Rev. B* **73** (2006) 155330.

18. P. Gu, M. Tani, S. Kono, K. Sakai, and X.-C. Zhang, "Study of terahertz radiation from InAs and InSb", *J. Appl. Phys.* **91(7)** (2001) 5533.

19. G. Klatt, F. Hilser, W. Qiao, M. Beck, R. Gebs, A. Bartels, K. Huska, U. Lemmer, G. Bastian, M. B. Johnston, M. Fischer, J. Faist, and T. Dekorsy, "Terahertz emission from lateral photo-Dember currents", *Opt. Exp.* **18(5)** (2010) 4939–4947.

20. G. Matthaus, S. Nolte, R. Hohmuth, M. Voitsch, W. Richter, B. Pradarutti, S. Riehemann, G. Notni, and A. Tunnermann, "Large-area microlens emitters for powerful THz emission", *Appl. Phys. B* **96(2)** (2009) 233–235.

21. R. Singh, E. Plum, C. Menzel, C. Rockstuhl, N. Zheludev, and W. Zhang, "Negative index in chiral metamaterials", *IEEE Photonic Society 24th Annual Meeting*, Arlington, VA, USA, Oct 2011, TuG1, 240–241.

22. D. V. Lavrukhin, A. E. Yachmenev, R. R. Galiev, R. A. Khabibullin, D. S. Ponomarev, Yu. V. Fedorov, and P. P. Maltsev, "MHEMT with a power-gain cut-off frequency of f_{max} = 0.63 THz on the basis of a In$_{0.42}$Al$_{0.58}$As/In$_{0.42}$Ga$_{0.58}$As/In$_{0.42}$Al$_{0.58}$As/GaAs nanohetero-structure", *Semiconductors* **48(1)** (2014) 69–72.

23. D. V. Lavrukhin, A. E. Yachmenev, R. R. Galiev, A. S. Bugaev, Y. V. Fedorov, R. A. Khabibullin, D. S. Ponomarev, and P. P. Maltsev, "Investigation and fabrication of the semiconductor devices based on metamorphic InAlAs/InGaAs/InAlAs nanoheterostructures for THz applications", *Int. J. High Speed Electron. Syst.* **24(1&2)** (2015) 1520001.

24. G. B. Galiev, R. A. Khabibullin, D. S. Ponomarev, A. E. Yachmenev, A. S. Bugaev, and P. P. Maltsev, "Metamorphic nanoheterostructures for millimeter-wave electronics", *Nanotechnol. Rus.* **10(7)** (2015) 593–599.

25. E. A. P. Prieto, S. A. B. Vizcara, A. S. Somintac, A. A. Salvador, E. S. Estacio, C. T. Que, K. Yamamoto, and M. Tani, "Terahertz emission enhancement in low-temperature-grown GaAs with an n-GaAs buffer in reflection and transmission excitation geometries", *J. Opt. Soc. Am. B* **31(2)** (2014) 291.

26. I. E. Ilyakov, G. Kh. Kitaeva, B. V. Shishkin, and R. A. Akhmedzhanov, "Terahertz time-domain electro-optic measurements by femtosecond laser pulses with an edge-cut spectrum", *Opt. Lett.* **41(13)** (2016) 2998–3001.

27. G. B. Galiev, S. S. Pushkarev, E. A. Klimov, P. P. Maltsev, R. M. Imamov, and I. A. Subbotin, "X-Ray diffractometry of metamorphic nanoheterostructures", *Cryst. Rep.* **59(2)** (2014) 258–265.

28. V. Apostolopoulos and M. E. Barnes, "THz emitters based on the photo-Dember effect", *J. Phys. D: Appl. Phys.* **47(37)** (2014) 374002.

Numerical Characterization of Dyakonov-Shur Instability in Gated Two-Dimensional Electron Systems

Akira Satou

Research Institute of Electrical Communications, Tohoku University,
2-1-1 Karahira, Aoba-ku, Sendai, Miyagi 980-8577, Japan
a-satou@riec.tohoku.ac.jp

Koichi Narahara

Department of Electrical and Electronic Engineering, Kanagawa Institute of Technology,
1030 Shimoogino, Atsugi, Kanagawa 243-0292, Japan
narahara@ele.kanagawa-it.ac.jp

We numerically analyze the system based on the essentially non-oscillatory shock capturing scheme in order to characterize the Dyakonov-Shur (DS) instability in a gated two-dimensional electron gas system (2DES). The predictions of the linearized model are examined for a 2DES sandwiched by the top and back metallic gates. By solving Poisson equation self-consistently, the dispersive properties of plasma wave are properly estimated. Special attention is paid to the impact of dispersion to nonlinear dynamics of plasma-wave oscillation. A single-gated 2DES is also investigated for demonstrating the DS instability in practical devices.

Keywords: two-dimensional plasma wave; terahertz; hydrodynamic transport equations.

1. Introduction

Plasma waves in a two-dimensional electron gas system (2DES) is known to exhibit instability caused by a cavity resonance of Doppler-shifted waves, called the Dyakonov-Shur (DS) instability.[1] Several investigations on the generation of far-infrared electromagnetic waves have been reported, based on the DS instability.[2-4] It requires asymmetrical boundary conditions, i.e., electron density at one end has to be kept fixed, while the other end must make electron momentum fixed. Then, the linearized model, reviewed briefly in Sec. II, predicts that the subsonic plasma wave can gain amplitude when electrons flows out at the momentum-fixed boundary. The model assumes plasma waves in a gated 2DES are dispersionless and the variations of electron density/ momentum density are sufficiently small for the waves not be suffered from nonlinearity significantly. In fact, the gates affect the density fluctuations in a 2DES differently when the spatial scales vary, such that finite wavelength dispersion is given to the electronic plasma wave in a 2DES. Moreover, the amplified variations of electron density inevitably violate the assumption that they stay small. At present, several practical devices that employ the DS instability as the key operation principle have been proposed. The device structure is sometimes too complicated to be described properly by the simplified model. It is thus very important to solve the practical potential spatio-temporal distribution with appropriate introduction of nonlinear dynamics.

In order to examine the DS instability in practical situations, we develop a numerical tool that solves the hydrodynamic transport equations of electron density and its momentum density with the Poisson equation based on the third-order essentially non-oscillatory shock capturing scheme. Although the hydrodynamic model simplifies the 2DES significantly in comparison to the kinetic-transport[5] and the Monte-Carlo models,[6] it shares the basic equations to be solved with the original linearized model discussed in Ref. 1; therefore, it has an advantage that the influences of both nonlinearity and dispersion on plasma waves can be observed in a distinct manner by the comparison with the predictions of the linearized model. The transport equations are solved in one dimension, such that they follow the one-dimensional motion of electrons in 2DES. Unlike the previous approaches,[7,8] the Poisson equations is solved in two dimensions for introducing the contribution of the near-by metal electrodes. By solving the two-dimension Poisson problem, we are free from imposing the long-wavelength approximations assumed in the linearized model. Moreover, the nonlinearity included in the transport equations is fully introduced, so that it becomes possible to examine the dynamics of plasma waves having large amplitude.

Keeping the potential and electron momentum fixed at one and the other ends, respectively, the numerical model simulates the situation that exhibits the DS instability. We first carried out several calculations for a 2DES with such boundary conditions that is sandwiched with the top and bottom gate electrodes. The two-side gates are required to suppress short-wavelength variations, which is prohibited to develop within the framework of the linearized model. A self-oscillation is successfully observed, whose dependence of frequency and growth rate on electron drift velocity is well characterized by the linearized model. It is shown that the quarter-wavelength resonance that corresponds to the lowest-order mode supports the DS oscillation for all the calculated cases with gain. As the electron density increases, the oscillation saturates to a limit cycle. In addition, the finite dispersion of plasma wave is shown to compensate for nonlinearity to develop solitonic waves.

Practically, two-sided-gate structure is difficult to be implemented. When only a single gate is set near the 2DES, the short-wavelength plasma waves cannot be completely suppressed. Such short-wavelength components can develop higher-order resonant modes, which interact non-trivially owing to nonlinearity. In order to examine the steady-state oscillation in such a situation we calculate the dynamics of electrons in a 2DES only with the top gate. It is established that the single gate suffices to excite a self-oscillation. However, it is supported by the third-order resonant mode mixed with the fundamental quarter-wavelength resonance, resulting in the spectral broadening.

2. Fundamental Properties of Dyakonov-Shur Instability

In this section, we briefly review the fundamental properties of the DS instability that are used for validating numerical results we obtain. We consider a 2DES with near-by metal gate electrode(s). We assume that both the 2DES and gates are infinitely thin. A Cartesian coordinate is set, such that z axis is normal to the 2D objects. The system is assumed invariant in y direction, i.e., we assume that the electrons in the 2DES can move only to the x direction. The continuity equation and the equation of motion are respectively given by:

$$\frac{\partial n}{\partial t} + \frac{\partial nv}{\partial x} = 0, \tag{1}$$

$$\frac{\partial v}{\partial t} + v\frac{\partial v}{\partial x} = -\frac{e}{m_e}\frac{\partial \varphi}{\partial x}, \tag{2}$$

where n, v, and φ show the electron density, velocity, and potential, respectively. Moreover, m_e shows the effective electron mass. The potential satisfies the Poisson equation

$$\frac{\partial^2 \varphi}{\partial x^2} + \frac{\partial^2 \varphi}{\partial z^2} = \frac{en}{\varepsilon_b}, \tag{3}$$

where ε_b represents the dielectric constant of the background material. When the 2DES is sandwiched by two metal electrodes as shown in Fig. 1(a), the Poisson equation is solved in the long-wavelength approximation to give

$$\varphi = \frac{ed_1d_2}{\varepsilon_b(d_1+d_2)}n + \frac{ed_1^2d_2^2}{3\varepsilon_b(d_1+d_2)}\frac{\partial^2 n}{\partial x^2}, \tag{4}$$

where d_1 (d_2) represents the separation between the top (bottom) gate and 2DES. Equations (1) and (2) are then closed for two variables n and v that are used to develop the linearized model of the DS instability.

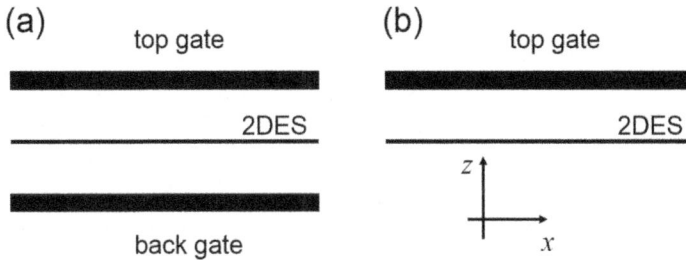

Fig. 1. Gated two-dimensional electron gas system (2DES). (a) The 2DES is sandwiched by the top and bottom metallic gates and (b) only the top gate is located above the 2DES.

For that instability, the electrode potential is fixed at φ_0 to guarantee the proper electron density in the 2DES. For sufficiently long wavelengths, the second term in the

R.H.S. of Eq. (4) can be neglected to give the plasma wave having the angular frequency ω and the wave number k satisfies the dispersion-free relationship $\omega = sk$, where the phase velocity s is given by $\sqrt{e\varphi_0/m_e}$ for the quiescent potential φ_0. For 2DES electrons drifting at the velocity v0, the plasma wave must be Doppler shifted for the dispersion to become $\omega = (s \pm v_0)k$, where the positive (negative) sign corresponds to the co-(anti-) moving electrons. For causing the DS instability, one of the boundaries of the 2DES have to set to a fixed potential and the current density at the other boundary have to be kept constant. We set the potential-fixed boundary at $x = 0$ and the current-fixed boundary at $x = L$. Then, the linearized model predicts that the plasma wave having the angular frequency of

$$\omega_0 = \frac{l\pi \left| s^2 - v_0^2 \right|}{2sL},\tag{5}$$

can be amplified with the growth rate of

$$\alpha = \frac{s^2 - v_0^2}{2sL} \ln \left| \frac{s + v_0}{s - v_0} \right|,\tag{6}$$

where l is a natural number that identifies the cavity resonance order. Moreover, φ is approximately given by

$$\varphi \propto \cos\left(\frac{l\pi}{2L} x \right) \sin\left(\frac{l\pi}{2sL} \left\{ \left(s^2 - v_0^2 \right) t + v_0 x + \phi_0 \right\} \right),\tag{7}$$

for the lth order plasma wave. For $|v_0| \ll s$, x dependence of the traveling-wave part of Eq. (7) can be neglected; therefore, the plasma wave operates like the standing wave with the wavelength of $4L/l$.

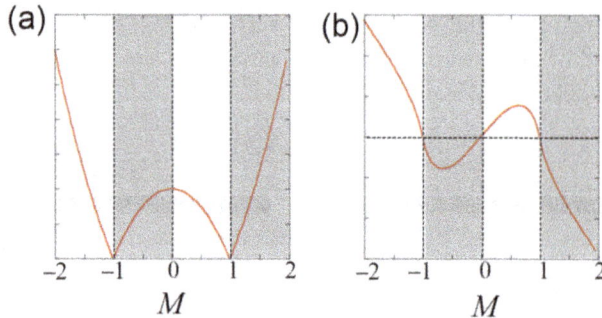

Fig. 2. The properties of the plasma wave exhibiting the DS instability. According to Eqs. (5) and (6), the dependence of ω_0 and α on the Mach number M is shown in Figs. 2(a) and 2(b), respectively. The ranges of M corresponding to negative α are hatched, where the plasma wave decays. The numerical values are omitted for the vertical axes for clarity.

For illustration, these properties of the plasma wave exhibiting the DS instability are shown in Fig. 2. Figure 2(a) shows the dependence of ω_0 on the Mach number defined by v_0/s when s is regarded as the sound velocity. For $|v_0| < s$, the frequency monotonically decreases and becomes zero at $|M| = 1$. Figure 2(b) shows the M dependence of α. For $v_0 > 0$, the plasma wave gains amplitude when it is subsonic, i.e., $|M| < 1$, while the subsonic plasma wave decays for $v_0 < 0$. The variation of the electron density becomes large for the amplifying plasma wave, such that the nonlinearity is expected to saturate the temporal variations. The linearized model does not guarantee such saturation mechanism. Moreover, it predicts nothing about the resonant mode the instability is supported by.

3. Numerical Model

We numerically solve the electron transport equations using the finite difference time-domain calculations with the essentially non-oscillatory (ENO) scheme[9,10] to examine the nonlinear properties of plasma waves. The electrical force that confines electrons to 2DES in z direction is strong enough that it is not always necessary to obtain exact electron transport in z direction. Thus, we only consider the electron transport in the x direction and set the equations to be solved as follows:

$$\frac{\partial n}{\partial t} = -\frac{\partial}{\partial x}\left(\frac{p}{m_e}\right), \tag{8}$$

$$\frac{\partial p}{\partial t} = -\frac{\partial}{\partial x}\left(\frac{p^2}{m_e n} + k_B T n\right) - e\frac{\partial \varphi}{\partial x} + \frac{p}{\tau_p}, \tag{9}$$

$$\frac{\partial^2 \varphi}{\partial x^2} + \frac{\partial^2 \varphi}{\partial z^2} = \frac{en}{\varepsilon_b}, \tag{10}$$

where p is the electron momentum density. The Poisson equation is solved in the two-dimensional space without using the long-wavelength approximation. Figure 3 shows the calculation setup. The spatial increment is 2.5 nm for both x and z directions, such that the total cell size is 70×20. In addition, $\varepsilon_b = 12.5$, $m_e = 0.07 m_0$, $T = 300$ K, $V_{G1} = V_{G2} = 0.05$ V, $d_1 = 10$ nm, and $d_2 = 12.5$ nm. The third-order ENO is used. The absorbing boundaries using ghost grids are set for both n and p. We employ Lax-Friedrichs numerical flux.[9] In order to simulate the boundary conditions to cause the DS instability, the left-end cell of the 2DES is set to be Dirichlet for the Poisson solver, while the electron momentum density at the right-end cell p_r is kept constant, whose value is given initially. For the Poisson solver, the boundaries are Neumann except for the above-mentioned left-end cell of the 2DES. The metallic gates are also set to be Dirichlet as shown by solid cells in Fig. 3. We first set p_r to zero and obtain the time-invariant steady-state solution of Eqs. (8), (9), and (10), which is employed as the initial values for the calculation for p_r to be nonzero. In order to simulate the predictions of the linearized model summarized in the previous section, τ_p is sufficiently large not to influence the dynamics unless stated. The initial time-invariant and spatially uniform electron density n_0 is 5.07×10^{11} cm^{-2}. The numerical sound velocity s_{num} is then estimated to be 4.8×10^7 cm/s by equating φ_0 with $V_{G1} + d_1 d_2 n_0/\varepsilon_b(d_1+d_2)$.

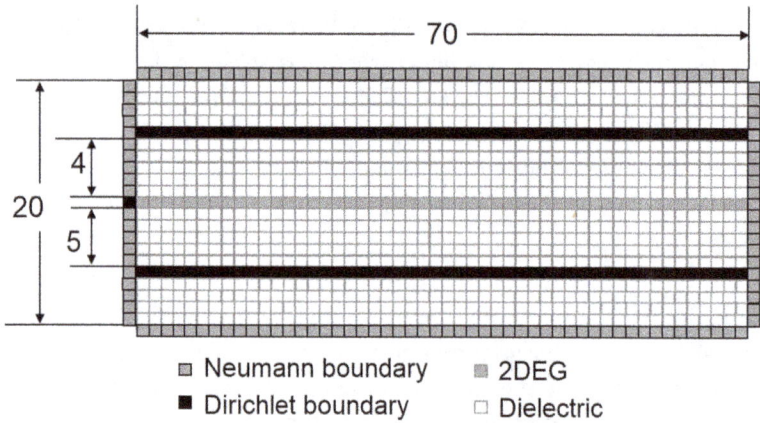

Fig. 3. The properties of the plasma wave exhibiting the DS instability. According to Eqs. (5) and (6), the dependence of ω_0 and α on the Mach number M is shown in Figs. 2(a) and 2(b), respectively. The ranges of M corresponding to negative α are hatched, where the plasma wave decays. The numerical values are omitted for the vertical axes for clarity.

4. Numerical Results

Figure 4 shows the calculated results obtained for $p_r = 1.0\times10^{-11}$ gs^{-1}cm^{-1}. Figure 4(a) shows the temporal variation of the electron density at $x = 100$ nm. The amplitude exhibits the exponential growth for the initial 8 ps and then saturates. It is thus established that the DS instability can lead to a limit-cycle oscillation of plasma wave, which can exhibit synchronization phenomena such as injection locking. Figure 4(b) shows the spatio-temporal waveform of the electron density corresponding to the saturated temporal variation. A standing wave whose unique node and anti-node exist at $x = 0$ and $x = 175$ nm, respectively. According to Eq. (7), the steady-state oscillation is shown to correspond to the $l = 1$ mode having the wavelength of $4L$.

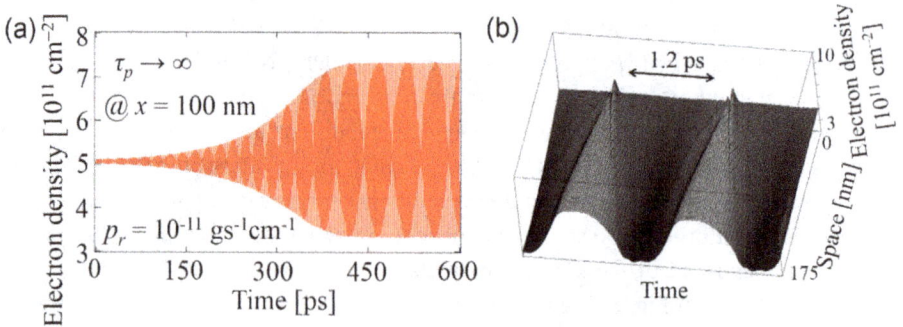

Fig. 4. The plasma wave gained by the DS instability. (a) The temporal variation monitored at $x = 100$ nm and (b) the steady-state spatio-temporal waveform.

In order to examine the properties predicted by the linearized model, we calculate the temporal evolution of the plasma wave for several different values of p_r. The average electron velocity is almost proportional to p_r, so that the M dependence shown in Fig. 2

should explain the results shown in Fig. 5. Figure 5(a) shows the oscillation frequencies for eleven positive p_r. Circles and squares represent the numerical and analytical frequencies, respectively. For analytical ones, Eq. (5) is estimated with s_{num} and numerically obtained v_0. The numerical oscillation frequency is smaller than the analytical one for all estimated p_r. Moreover, it decreases as p r becomes large at much greater extent than the analytical one. In numerical calculations, the variation of electron density naturally grows at positive p_r, so that it cannot stay small enough to validate the linearized model. The discrepancy in frequency is supposed to be mainly caused by this large density variation. The short-wavelength dispersion is also shown to influence the plasma oscillation frequency by solving the Poisson equation using Green function.[11] In addition, ω_0 is very sensitive to s_{num} (ω_0 becomes $\approx$ 600 GHz when s_{num} is reduced slightly to 4.3×10^7 cm/s). On the other hand, Fig. 5(b) shows the growth rate for both positive and negative p_r. Circles represent the numerically obtained growth rates, while the analytical rates obtained by estimating Eq. (6) are shown by squares. The coincidence establishes the fact that the two different methods well characterize the small plasma wave oscillations based on the DS instability.

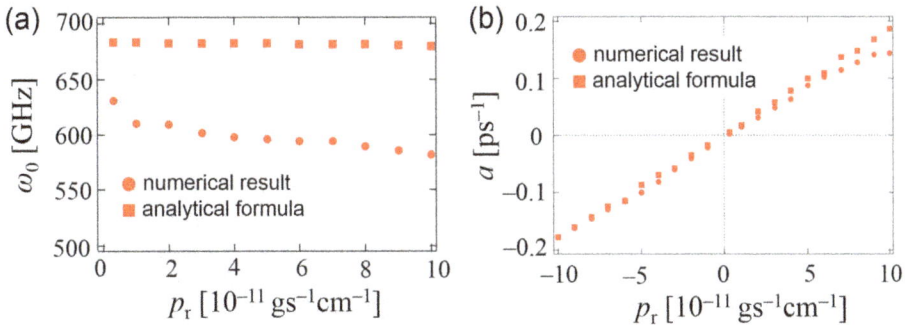

Fig. 5. Numerically obtained properties of DS instability. The dependence of (a) the oscillation frequency and (b) the growth rate on the electron momentum density at the right end of the 2DES are shown. Squares and circles show analytical and numerical dependences, respectively.

The amplitude dependence of propagation property is examined. Several spatial waveforms for plasma-wave oscillations are plotted at 150 fs intervals in Fig. 6. Both the small and the large amplitude waveforms are shown for comparison. At the small amplitude shown in Fig. 6(a), the electron density decreases or increases from the node at $x = 0$ to the antinode at $x = 175$ nm to exhibit the typical quarter wavelength resonance. In Fig. 6 (b), the wave property for large amplitude is plotted at 150 fs intervals. Govorov *et al.*[12] discussed the development of electronic solitons in a 2DES with both the top and back gate metals. The balance between the nonlinear terms present in Eqs. (8) and (9) and the third-order dispersion results in developing solitonic waves.[12-16] In Ref. 7, the nonlinearity results in the shock formation because the plasma waves are assumed to be dispersionless. It is thus essential for the Poisson equation to be coupled without any approximations for proper description of nonlinear plasma waves. The excitation of electronic solitons means that the localized electron bunch can travel in a 2DES without distortion. The velocity of solitons is expected to be larger than the diffusion/drift electron velocity; therefore, the solitons are supposed to trigger effectively the Smith-Purcell effect in solid-state devices. Such solitary waves are observed in solid waveforms in Fig. 6(b), so that each solitary wave gains amplitude as its propagation, although further investigation must be carried out to examine their practical importance, including the influences of finite momentum relaxation time and the grating structure that effectively couples with electronic solitons.

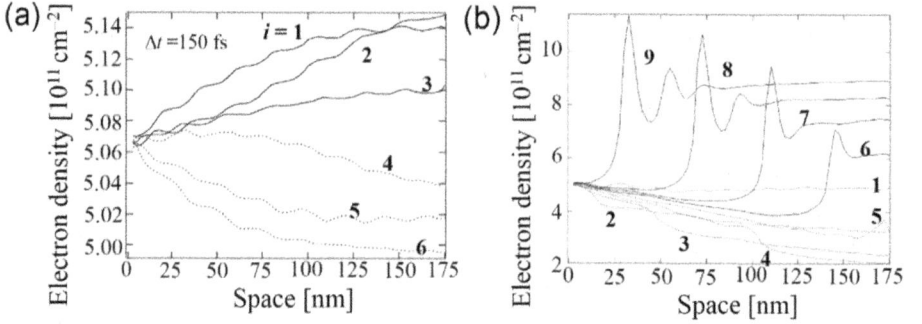

Fig. 6. Solitonic waves observed in large plasma oscillation. The spatial waveforms recorded at 150 fs intervals (numbered as i = 1, 2, 3...) are shown for (a) small and (b) large plasma oscillation. The waveforms for the electron density to be above and below its average at the right end (x = 175 nm) are shown by the solid and dotted curves, respectively.

In general, the back gate is difficult to be implemented in actual devices. We thus consider the case when an electrode is uniquely present near the 2DES as shown in Fig. 1(b). The Poisson equation is then solved in the long-wavelength approximation as above to give

$$\varphi = \frac{ed}{\varepsilon_b}n - \frac{2ed^3}{3\varepsilon_b}\frac{\partial^2 n}{\partial x^2} + \frac{2ed^2}{\varepsilon_b}\frac{\partial}{\partial x}P.V.\int\frac{dx'}{2\pi}\frac{n(x+x')}{x'}, \tag{11}$$

where $P.V.$ stands for the principal value and d represents the separation between the top gate and 2DES. Although the last term in the R.H.S. is the second order in magnitude, it results from the non-local nature of the electron-electron interaction and can influence the dispersionless nature of the plasma wave non-trivially.

Figure 7 shows the calculated results for a plasma wave in a single-gated 2DES. Only the back gate is eliminated in Fig. 3. The gate potential V_{G1} is set to 0.08 V for rendering equivalent electron density to the two-sided-gate cases. p_r is set to 0.3×10^{-11} gs^{-1}cm^{-1} and τ_p is set to 10.0 ps that is the value achievable at low temperatures. Figure 7(a) shows the temporal variation in electron density monitored at x = 100 nm. The DS instability is successfully triggered to gain plasma-wave amplitude. However, the mono-cycle nature established in the two-sided-gate cases is lost. The resonance dynamics at around 6 ps become complicated, and then relaxes to a steady state, whose spatio-temporal waveform is shown in Fig. 7(b). Owing to the ungated portion, the self-induced short-wavelength variations excite the higher-order resonances together with the fundamental one. For the present case, the 3rd order resonance dominantly coexists to result in the quasi-periodic oscillation. Influences caused by the short-wavelength interactions are not exhausted. However, the dynamics tend to relax to a steady quasi-periodic state only a few resonant modes coexist after experiencing complicated process relating with many higher-order resonances. As a result, the elimination of the back gate leads to the spectral broadening by the presence of several different resonance modes.

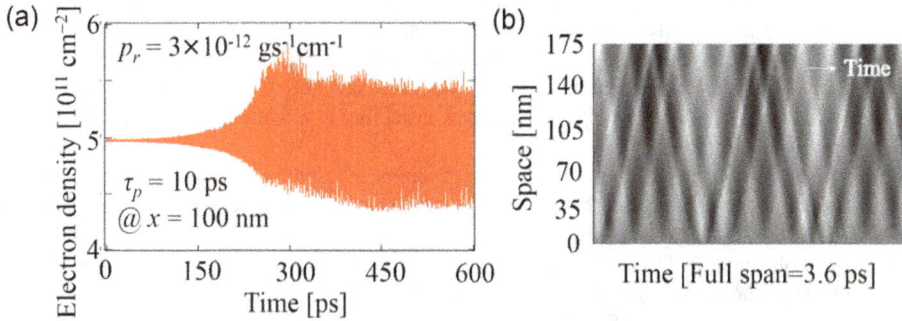

Fig. 7. Properties of DS instability in single-gated 2DES. (a) The temporal variation of electron density monitored at $x = 100$ nm and (b) the spatio-temporal profile of steady-state electron density.

5. Conclusion

The DS instability in the two-sided-gated 2DES is investigated by ENO-based numerical calculations. The dependence of the oscillation frequency and the growth rate on the electron velocity is estimated to be coincident with the predictions of the linearized model. The calculations show that the nonlinearity saturates the plasma oscillation to result in a limit cycle and suggests that a quarter-wavelength resonance dominates the two-sided-gate 2DES with positive electron velocities. On the other hand, the DS instability in a single-gated 2DES leads to multiple excitations of resonance modes of different order. The complicated dynamics observed in the single-gated 2DES signify that the numerical methods would occupy very important position for characterizing practical oscillators based on the DS instability.

Acknowledgments

Part of this work was carried out under the Cooperative Research Project Program of the Research Institute of Electrical Communication, Tohoku University.

References

1. M. Dyakonov and M. S. Shur, "Shallow water analogy for a ballistic field effect transistors: new mechanism of plasma wave generation by dc current", *Phys. Rev. Lett.* **71** (1993) 2465–2468.
2. V. V. Popov, G. M. Tsymbalov, and M. S. Shur, "Plasma wave instability and amplification of terahertz radiation in field-effect-transistor arrays", *J. Phys.: Condens. Matter.* **20** (2008) 384208.

3. Y. M. Meziani, T. Otsuji, M. Hanabe, T. Ishibashi, T. Uno, and E. Sano, "Room temperature generation of terahertz radiation from a grating-bicoupled plasmon-resonant emitter: size effect", *Appl. Phys. Lett.* **90** (2007) 61105.
4. V. Ryzhii, A. Satou, and M. S. Shur, "Plasma instability and terahertz generation in HEMTs due to electron transit-time effect", *IEICE Trans. Electron.* **E89-C** (2006) 1012–1019.
5. A. Satou, V. Ryzhii, V. Mitin, and N. Vagidov, "Damping of plasma waves in two-dimensional electron systems due to contacts", *Phys. Status Solidi B* **246** (2009) 2146–2149.
6. J.-F. Millithaler, L. Reggiani, J. Pousset, L. Varani, C. Palermo, W. Knap, J. Mateos, T. Gon´zalez, S. Perez, and D. Pardo, "Monte Carlo investigation of terahertz plasma oscillations in ultrathin layers of n-type $In_{0.53}Ga_{0.47}As$", *Appl. Phys. Lett.* **92** (2008) 042113.
7. A. P. Dmitriev, A. S. Furman, V. Y. Kachorovskii, G. G. Samsonidze, and Ge. G. Samsonidze, "Numerical study of the current instability in a two-dimensional electron fluid", *Phys. Rev. B* **55** (1997) 10319–10324.
8. S. Rudin, G. Samsonidze, and F. Crowne, "Nonlinear response of two-dimensional electron plasmas in the conduction channels of field effect transistor structures", *J. Appl. Phys.* **86** (1999) 2083–2088.
9. C.-W. Shu and S. Osher, "Efficient implementation of essentially non-oscillatory shock-capturing schemes, II", *J. Comp. Phys.* **83** (1989) 32–78.
10. E. Fetami, J. Jerome, and S. Osher, "Solution of the hydrodynamic device model using high-order nonoscillatory shock capturing algorithms", *IEEE Trans. CAD* **10** (1991) 232–244.
11. A. Satou, I. Khmyrova, A. Chaplik, V. Ryzhii, and M. Shur, "Spectrum of plasma oscillations in slot diode with two-dimensional electron channel", *Jpn. J. Appl. Phys.* **44** (2005) 2592–2595.
12. A. O. Govorov, V. M. Kovalev, and A. V. Chaplik, "Solitons in semiconductor microstructures with a two-dimensional electron gas", *JETP Lett.* **70** (1999) 488–490.
13. A. V. Chaplik, "Possible crystallization of charge carriers in low-density inversion layers", *Sov. Phys. JETP* **35** (1972) 395–398.
14. E.Vostrikova, A. Ivanov, I. Semenikhin, and V. Ryzhii, "Electrical excitation of shock and solitonlike waves in two-dimensional electron channels", *Phys. Rev. B* **76** (2007) 035401-1–8.
15. I. Semenikhin, V. Ryzhii, E. Vostrikova, and A. Ivanov, "Electrical excitation of shock and soliton-like waves in high-electron-mobility transistor structures", *Phys. Stat. Sol. (c)* **5** (2008) 61–65.
16. K. Narahara and Y. Suzuki, "Characterization of plasma waves in gated two-dimensional electron systems", *J. Appl. Phys.* **103** (2008) 023301.

Author Index